INTERACTION AND
COEVOLUTION

INTERACTION AND COEVOLUTION

JOHN N. THOMPSON

Washington State University

A Wiley-Interscience Publication

JOHN WILEY & SONS

New York   Chichester   Brisbane   Toronto   Singapore

Library of Congress Cataloging in Publication Data:

Thompson, John N.
    Interaction and coevolution.

    "A Wiley-Interscience publication."
    Includes bibliographical references and index.
    1. Evolution. I. Title. II. Title: Coevolution.
QH371.T49 1982   575   82-11026
ISBN 0-471-09022-0

Printed in the United States of America

10 9 8 7 6 5 4 3 2 1

PREFACE

Species interact in a bewildering variety of ways and the evolutionary results of these interactions are stamped in their morphologies, physiologies, behaviors, and life histories. It is not only the species that change evolutionarily through interactions, however; the interactions themselves also change. My intent in writing *Interaction and Coevolution* is to synthesize general patterns in the evolution of interactions and in the conditions likely to favor coevolution among interacting species. The emphasis throughout is on patterns that transcend taxonomic boundaries—that is, on patterns that derive more from the mode of interaction than from the biology of a particular taxonomic group. The conclusions in each chapter should be read as hypotheses in need of thorough testing. In some cases the conclusions rest on a solid empirical base, whereas in other cases the conclusions are derived from a synthesis of a diverse literature in which the data have been gathered for a wide variety of reasons. In either circumstance the hypotheses provide a framework within which to collect, analyze, and compare future ecological data on the evolution of interactions.

I have been fortunate while writing this book to have colleagues who gave generously of their time to listen to, argue about, or comment on viewpoints and who shared their own ideas through comments or preprints. I am particularly grateful to Peter W. Price for his comments on the original outline and all the completed chapters and to Mary F. Willson for her critique of the outline and the chapters on mutualism. Their thoughtful and candid comments on the early drafts of some chapters greatly helped me to focus my arguments. I also thank John F. Addicott, Douglas W. Biedenweg, Susan Stone Bookman, Lawrence E. Gilbert, Daniel H. Janzen, Richard N. Mack, Steward T. A. Pickett, John Kingsbury Piper, Douglas W. Schemske, Nathaniel Wheelwright, Thomas G. Whitham, and David Sloan Wilson for reading chapters or discussing ideas on the evolution of interactions with me. Each of them made comments that influenced my ideas as I was writing the book. Frank Crome, Douglas Schemske, Nathaniel Wheelwright, and David Wilson kindly sent unpublished data or manuscripts.

I also thank Mary M. Conway for help in all aspects of editing the book, from its initial conception to its completion. Cindy Craver cheerfully typed the several drafts of the manuscript. Finally, I thank my wife, Jill, who has been willing to listen and discuss over dinner the earliest inklings of ideas that appear in this book. I dedicate this book to her.

JOHN N. THOMPSON

Pullman, Washington
August 1982

CONTENTS

INTERACTION AND COEVOLUTION

CHAPTER

1

EVOLUTIONARY ECOLOGY, INTERACTION, AND COEVOLUTION

As species have evolved and multiplied over the past three and a half billion years, so have the interactions that link their life histories and shape their further evolution. Although interactions are often more ephemeral and certainly less tangible than species, they are as much a product of evolution as is bone, chitin, or cell wall: biological communities differ from zoos and botanical gardens because of the interactions between species. Therefore, the study of the origins and evolution of interactions between species is as crucial to an understanding of the evolution of life as is the study of the origins and evolution of species.

We know much about the effects of interspecific interactions on the life histories, morphologies, and behaviors of organisms and on the size, structure, and dynamics of populations. However, we lack an overall conceptual framework for the evolution of interactions that could suggest general patterns in the ways organisms respond evolutionarily to interactions and the ways interactions change over evolutionary time. If we can identify general patterns, then we can better understand the constraints that different kinds of interaction impose on the evolution of organisms and the changes that occur in the interaction structure of communities over evolutionary time.

My purpose in this book is toward a general theory for the evolution of interspecific interactions. Specifically, I focus on the following set of questions:

1

1. Are there general patterns in how interactions differ in the selection pressures that they exert on organisms?
2. Are there general patterns in how interactions are likely to change in their effects on the fitness of organisms over evolutionary time?
3. Under what ecological conditions is coevolution likely among interacting species?

That is, how do interactions differ evolutionarily, how do they change, and when do they generate reciprocal change among species?

Chapters 2 and 3 are devoted to the evolution of antagonistic interactions. I begin in Chapter 2 by considering differences in patterns of specialization, defense, and coevolution between parasites, grazers, and predators and their victims. This chapter emphasizes that the mode of interaction between species is critical to understanding how selection acts on interactions and when coevolution is likely. The conclusions of this chapter are carried through the remainder of the book. In Chapter 3 I contrast competition with other kinds of antagonistic interaction and consider the conditions under which long-term coevolution is likely between competitors.

Chapters 4, 5, and 6 concern the evolution of mutualisms. In Chapter 4 I discuss the close evolutionary relationship between antagonism and mutualism and argue that the commonness of mutualisms in communities often depends evolutionarily on the richness of antagonistic interactions. In Chapter 5 I consider general and specific aspects of life histories that favor the evolution of mutualisms. These two chapters provide the background for the range of ecological conditions that are likely to favor mutualisms, and Chapter 6 considers the subset of mutualisms that are likely to generate coevolution among mutualists.

Chapter 7 is a departure from the flow of argument on coadaptation in Chapters 2 through 6. This chapter considers the effects of different kinds of interaction on patterns of speciation and on the likelihood of cospeciation resulting from interactions. I differentiate cospeciation from simple phylogenetic tracking of one taxon on another. I argue also that the tempo of speciation may differ between species involved in the same interaction and this may influence importantly our views on the commonness of cospeciation.

Finally, Chapter 8 concentrates on a group of questions that form the basis for the study of the interaction structure of communities and for the importance of coevolution in structuring interactions within communities. These questions concern the pattern of development and change in interactions within and among communities in contemporary time—the patch dynamics of interactions—and the growth and change in interactions over evolutionary time through coadaptation, cospeciation, and collection of unrelated species into interactions.

Together the arguments in these chapters are an attempt toward a general framework within which to study the evolution of interactions, the likelihood of coevolution among interactions, and the interaction structure of communities. I emphasize throughout the conditions under which coevolution is likely or unlikely in the evolution of interactions. Natural communities are not superorganisms, but they are also not random collections of species with no evolutionary effects on each other. The study of patterns in coevolution can help us to understand where and when interactions will bind the gene pools of two or more species in a community through reciprocal evolutionary change.

THE COEVOLUTIONARY PERSPECTIVE

Coevolution is reciprocal evolutionary change in interacting species. The key word is reciprocal in the sense of mutual. In an interaction between two species, both species must undergo evolutionary change specifically in response to the interaction to be called a coevolved interaction. As Janzen (1980a) notes, simply showing mutual congruence of traits between a pair of species does not necessarily indicate coevolution has occurred. Upon entering a new habitat, a species will tend to interact with other species whose traits are most congruent with its own, as in phytophagous insects that colonize introduced plants (e.g. Strong, 1974, 1979). No coevolution may have occurred between the species in this instance. The concept of coevolution is a powerful tool for evolutionary theory and the use of the term coevolution, as well as its sister terms coadaptation and cospeciation, should be used carefully to describe only interactions where the evidence suggests that reciprocal evolution has occurred.

The other key part of the definition as given here is that coevolution involves the partial coordination of nonmixing gene pools. That is, it is an interspecific process. Although this is the way the terms coevolution and coadaptation are used mostly in evolutionary ecology, these terms have been used in other subdisciplines of evolutionary theory to describe coordinated evolution at lower levels in the hierarchical organization of life. Even Darwin used the term coadaptation in two senses in the *Origin*. In the introduction he wrote of the "coadaptations of organic beings to each other and to their physical conditions of life." The first usage is the sense in which the term is used today in evolutionary ecology usually interchange-ably with the term coevolution. The second usage relates more to the coordinated development of parts within an organism rather than to reciprocal adaptations between species. In addition, Dobzhansky (1970 and earlier) used the phrase coadapted gene complexes in the specific sense of mutual adjustment of closely linked genes that function together within chromosomes, and that usage continues within the field of population genetics (Lewontin, 1974; Roughgarden, 1979). At the level of molecular genetics, others have written of the coevolution of the genetic code (Lacey et al., 1975; Wong, 1975). Since there is no arbiter of terms in science, all of these uses are likely to continue side by side in the literature; in the following chapters coevolution (coadaptation) always means reciprocal change (adaptations) among interacting species.

THE DEVELOPMENT OF THE COEVOLUTIONARY PERSPECTIVE

The coevolutionary perspective has developed as a merger of several research orientations. Although a full history is not possible here, it is worthwhile to emphasize a few points to show the variety of research directions from which the perspective has developed. Ehrlich and Raven's (1964) study on coevolution in butterflies and plants was certainly the paper that established the coevolutionary perspective as a major framework within which to study the evolution of interactions. Brues (1924) made similar arguments earlier for interactions between insects and plants (see Gilbert, 1979), but his arguments deal more with host shifts in insects than with reciprocal evolution and were made before the science of evolutionary ecology had developed sufficiently to incorporate a coevolutionary approach to the study of interactions. When Ehrlich and Raven's paper appeared in 1964, however, evolutionary ecology was rapidly coalescing as a major branch of ecology. Lack's studies of birds were cast explicitly in the framework of evolutionary ecology and his 1965 presidential address to the British Ecological Society was entitled "Evolutionary Ecology." Hutchinson's (1965) book *The Ecological Theater and the Evolutionary Play* also reflects the development of this orientation in ecology as does Orians' (1962) paper entitled "Natural Selection and Ecological Theory." The study of coevolution fit comfortably within the developing field of evolutionary ecology.

At least two other research approaches were also leading toward the development of a coevolutionary perspective in the early 1960s. Flor's (1942, 1955) concept of gene-for-gene interactions between parasites and hosts, developed initially in the literature on phytopathology, was being introduced into the broader literature (e.g. Person et al., 1962). Mode's (1958) paper in *Evolution,* entitled "A mathematical model for the co-evolution of obligate parasites and their hosts," is probably the first mathematical model of the genetics of coevolution and is explicitly a model of Flor's results on gene-for-gene interactions in flax and flax rust.

In addition, Pimentel's studies of what he calls genetic feedback began appearing in 1961. These studies are aimed at understanding the relationship between coevolution and population regulation, although he does not use the term coevolution in his early papers. The logic of these experiments is as follows: organisms adapted to feeding on a host are able to achieve high population densities; these high densities create strong selection pressures on host populations and select for resistance to attack; this feeds back negatively on the enemy (predator or parasite) population; after many population cycles, enemy populations are ultimately limited and stability results (Pimentel, 1961). This effect has been shown to some extent in the housefly *Musca domestica* and its parasitoid *Nasonia vitripennis* when the parasite is interacting with resistant hosts as compared with susceptible hosts under laboratory conditions (Pimentel et al., 1978). The parasite populations on experimental (resistant) hosts oscillate around a lower mean, oscillate less in absolute numbers of individuals, and have an overall tendency toward decreased oscillation. This, of course, is only one facet of coevolution and under these conditions, the host is not freely evolving in response to the parasite from the start: the experiments were begun with populations of susceptible and resistant houseflies. Although the assumptions of the genetic-feedback models were highly restrictive (Lomnicki, 1971; Slatkin and Maynard Smith, 1979), the perspective of the experiments was nonetheless coevolutionary.

The study of coevolution, therefore, developed from a variety of research orientations and proceeded hand-in-hand with the development of evolutionary ecology as a major research framework. Janzen's (1966, 1967) landmark study of acacias and acacia ants illustrated how coevolution could be studied through a combination of solid natural history observation and experimentation within natural communities. By 1975 a variety of interactions between animals and plants could be summarized from a coevolutionary perspective in the important volume on *Coevolution of Animals and Plants* edited by Gilbert and Raven. Mathematical modeling of coevolution in two-species systems also became an important direction for modeling in population genetics and evolutionary ecology in the 1970s (Roughgarden, 1979; Slatkin and Maynard Smith, 1979).

Together these approaches moved the study of coevolution toward analyses of general patterns among particular kinds of interaction. For example, Price (1977, 1980) asks how coevolution between parasites and their hosts could differ from coevolution between predators and prey, and Connell (1980) suggests why coevolution may be infrequent among competitors. This book, then, is part of the search for patterns in coevolution and in the evolution of interactions in general.

2

PARASITISM, GRAZING, AND PREDATION

Antagonistic interactions occur between species because living organisms are concentrated packages of energy and nutrients (trophic interactions) and because resources are limited (competition). That species respond evolutionarily to these interactions is evident immediately; Darwin used antagonistic interactions more than any other kind to illustrate in the *Origin of Species* how selection works. The problem in evolutionary ecology, however, is not simply to catalog how species respond evolutionarily to antagonistic interactions but rather to discern patterns in how species respond.

Some evolutionary patterns may result from the ways organisms feed on other species. Organisms differ greatly in how they attack their victims, including whether they kill their victims, how long they remain to feed on a single victim before killing it or leaving it, and how many victims they feed upon during their lifetimes. These differences in modes of feeding influence how (1) organisms specialize on their victims, (2) victims defend themselves against enemies, and (3) coevolution proceeds between enemies and their victims. This chapter considers patterns in specialization, defense, and coevolution between parasites, grazers, and predators and their victims in an effort to find patterns in interaction and coevolution that transcend taxonomic boundaries.

MODES OF FEEDING

The number of ways of categorizing interactions between species is probably unlimited. Terms such as grazer, browser, predator, parasite, herbivore, and carnivore vary among researchers in the breadth or narrowness of their usage (Dindal, 1975; Starr, 1975). For example, Harper (1977) uses the term predation in a general way to describe all the different ways that herbivores feed on plants, whereas Lubchenko (1979) restricts the use of the term to consumers that kill their hosts. All classifications of interactions are necessarily artificial, and their functional purpose can be only as aids to grouping some of the differences in how species interact. Here I separate modes of feeding into parasitism, grazing, and predation because these categories seem useful in assessing some patterns in how selection acts on interacting species. These categories are no more discrete than the concepts of population or community, but like these latter concepts they provide a useful tool for comparative studies in ecology.

Parasitism

A parasite is an organism that lives throughout a major period of its life in or on a single host individual, deriving its food from the host and causing lowered survival or reproduction in the host. Although the term parasite is used most often with reference to small animals or microorganisms that live in or on larger animals—as suggested by the usual content of parasitology journals and texts—this way of obtaining food is common among many taxa. Parasites include, for example, parasitic fungi of plants, herbivores such as caterpillars and aphids that spend all of their larval or nymphal stages on a single host plant, insect parasitoids of other insects that live as larval parasites in their hosts but have a free living adult stage, as well as the parasites usually considered in parasitology such as the cestode parasites of animals. Including the interactions between some phytophagous insects and plants in the application of the term parasite does not broaden the use of the term beyond utility, as has been suggested by some authors (Holling, 1980; Brooks, 1981). Instead, it does just the opposite by using the term in an ecological and evolutionary sense rather than in a sense set arbitrarily by taxonomic boundaries. The purpose of the broad application of the term parasite is to search for general principles that transcend taxonomy in the interactions between organisms (Price, 1975b; 1977; 1980).

The lack of an absolute discrete boundary between parasitism and other kinds of interactions between victims and their enemies is at once both exciting to the evolutionary ecologist interested in the evolution of interactions between species and the transition states between, for example, parasitism and predation, and also frustrating for those interested in categorizing in stone how species interact. Wheeler (1911) notes that "parasitism is an extremely protean phenomenon, one which escapes through the meshes of any net of scholastic definitions in which we may endeavor to confine it." Using a definition essentially the same as the one given above, however, Price (1977) estimates that over 50% of extant species are parasitic.

Parasite–host interactions differ from grazer–host and predator–prey interactions in the following ways:

1. The parasite lives in intimate association in or on its host.
2. The association lasts for a major portion of the parasite's life.
3. All (or almost all) of the parasite's food is derived from one host.
4. Death of the host prior to the stage at which the parasite is adapted to leave the host generally results in death of the parasite.

Grazing

A grazer is an organism that feeds on parts of several to many hosts during its lifetime and causes some detrimental effects on fitness of its hosts excluding immediate death.

Grazers include an array of organisms with a wide variety of feeding habits. Grazers of plants are common especially among insects (e.g. grasshoppers and leaf-cutter ants). Grazers of animals include vampire bats and biting flies that feed on the blood of mammals, as well as starfish that graze on corals, and cleaner fish mimics that bite off bits of flesh from coral reef fish. Grazing on animals may be less common than grazing on plants because the nonmodular construction of most animal bodies often does not allow for the removal of parts without killing the animal. Most higher plants exhibit a reiterative pattern of growth (Bazzaz and Harper, 1977; Harper, 1977), and the removal of one or several of the iterations often does not kill the plant. A bite taken from the side of an animal, however, often results in the death of the animal and, therefore, is regarded as predation. There are, of course, some exceptions to this statement. Corals exhibit a colonial structure much closer to plants than to most animals. Except when a colony is small, feeding on corals often more closely parallels grazing on plants than predation on animals (Connell, 1973).

Grazer–host interactions differ from parasite–host and predator–prey interactions in the following ways:

1. The interaction lasts for a relatively short period of the grazer's life, unlike parasites but like predators.
2. The interaction generally does not result in death of the host, unlike predation and like many, but not all, instances of parasitism.
3. The grazer may interact with many hosts during its lifetime and

often within a single feeding bout, unlike parasites but like predators.

Predation

A predator is an organism that feeds on several to many individuals during its lifetime, quickly killing each individual on which it feeds. The ambiguous part of this definition, of course, is the adverb quickly. Quickly is important here in the evolutionary sense that the prey must respond immediately to an interaction with a predator in order to avoid being killed within the next few seconds to hours. (The prey may occasionally take hours to die as in moose wounded by wolves.) Therefore, predator–prey interactions differ from parasite–host and grazer–host interactions in the following respects:

1. By definition, predators kill their prey unlike many parasites or grazers.
2. A predator eats several to many prey during its lifetime, unlike parasites but like grazers.
3. Predators kill their prey quickly so the response time available to prey is short relative to the time hosts have available to respond to parasites.

Transitions

The categories of parasitism, grazing, and predation represent pure forms of the antagonistic interactions that occur between species at different trophic levels. An individual species may share characteristics of two or even all three of these categories, but developing concepts of the ideal forms of the categories of interaction can make clearer the evolutionary transitions among feeding modes. It is worthwhile following through one example before considering the effects of feeding modes on specialization, defense, and coevolution.

In *Fleas, Flukes and Cuckoos,* one of the classic works on parasitism, Rothschild and Clay (1952) assert that in the evolution of the myriad parasitic relationships, the only common element is opportunity. Nevertheless, there are some predictable evolutionary pathways from one mode of interaction to another, and species occur at different points along those pathways. A well-studied example is the close association of insects and vertebrates.

Close associations with vertebrates have arisen in seven insect orders:

Dermaptera, Phtiraptera (Mallophaga and Anoplura), Hemiptera, Coleoptera, Diptera, Siphonaptera, and Lepidoptera (Waage, 1979). These associations vary in the length of time that the insect and its vertebrate host are in contact and in the effect of the interaction on the host, ranging from parasitic to commensalistic to mutualistic. Taxa whose individuals are only briefly associated with a host, such as mosquitoes and bed bugs are essentially grazers in their evolutionary biology. The differences between these taxa and those with more intimate associations with vertebrate hosts seem to relate to differences in the evolutionary pathways that have led to the associations.

Waage (1979) suggests that the intimacy of ectoparasite–host relationships depends upon the sequence in which the parasite evolves adaptations to feed on the host. He divides the kinds of adaptations needed to develop such a relationship into two types: (1) those adaptations promoting regular contact with the host, including habitat preference, host-finding behavior, and morphological adaptations for living on the host; and (2) those adaptations promoting ability to feed on the host, including mouthpart structure, physiological ability to digest host tissues, and behavior associated with the initiation and termination of feeding.

Insect species evolving along pathway I (type 1 adaptations followed by type 2 adaptations) are generally those whose ancestors were associated with dung or with vertebrate dwellings such as nests, roosts, or burrows. Development of mouthparts, physiology, and behavior allowing for direct feeding on host blood or other tissues occur after those adaptations promoting prolonged intimate association with the host. This seems to be a common pathway to insect ectoparasitism of vertebrates and is found in the Mallophaga, Coleoptera, and some Dermaptera, Diptera, and Lepidoptera. Many of the associations evolved through this pathway, however, have not involved the secondary evolution of traits allowing the insect to feed directly on living host tissue and are not parasitic (Waage, 1979).

Pathway II (type 2 adaptations followed by type 1 adaptations) involves insect species with mouthpart structures preadapted to feeding on vertebrate tissues through skin piercing (Hemiptera, some Diptera, and Lepidoptera) or eye feeding (some Diptera and Lepidoptera). The pathway is from free-living insects not associated with vertebrates, to brief feeding bouts usually by adults, to gradual secondary acquisition of close associations with the host or its microhabitat (Waage, 1979). The associations of individuals with hosts, however, often remain brief.

In general, it seems that the truly parasitic relationships involving prolonged intimate association with the vertebrate host are more likely through pathway I, although only a fraction of those insects following this pathway actually become parasitic. The final evolutionary step of feeding on vertebrate tissue often involves a mutualistic relationship between the insect and gut microorganisms, especially if the host food is nitrogen-rich blood. Vertebrate blood is deficient in B vitamins and these are synthesized by the microorganisms in the gut of blood-feeding insects parasitic on vertebrates (Waage, 1979). Because evolution along these pathways is generally a multistep process and selection can favor organisms with any combination of traits along these pathways, clean distinctions between parasitism and grazing for all species are not expected. Jennings (1974) has published a similar kind of study for the Turbellaria (flatworms) focusing on the combination of morphological and physiological adaptations found in the evolutionary continuum between predation and parasitism among these species.

SPECIALIZATION ON HOSTS AND PREY

All organisms specialize. The dichotomy often created between specialists and generalists is artificial and derives from consideration of only one or two of the many resource axes along which organisms can specialize. The question of specialization is usually phrased as why some organisms are host specific whereas others are less host specific, but the question could also be phrased as why organisms specialize in one way rather than another. Therefore, the related question is not only whether host-specific organisms are more likely to coevolve with hosts (or victims or prey) than organisms that are less host specific, but also how different ways of specialization affect coevolution.

  The designation of organisms as specialists or generalists with respect to host or prey species and explanations of patterns of specialization suffer from three types of problems: semantic, taxonomic, and conceptual. Semantic problems result from the great variety of ways in which the terms specialist and generalist, or monophagous and polyphagous, are used. In some studies of specificity, for example, these terms are applied based on the number of higher taxa on which a species feeds rather than the number of host or prey species (Table 2.1). There is, of course, some logic behind this method of classifying feeding habits. Coadaptation of parasites and hosts, for example, may be independent of host speciation: a parasite feeding on two closely related hosts may be more tightly bound to its hosts' genomes than a parasite feeding on two hosts in different orders. But by following this logic, a parasite feeding on 12 hosts in the same host family is more specialized than a parasite with two hosts in different orders. Both analyses based on the number of species that serve as hosts and those based on the number of higher taxa utilized are legitimate and valuable but they

Table 2.1.
Some Definitions of Terms for the Diets of Organisms

Category	Specialist (S) or Monophagous (M)	Generalist (G) or Polyphagous (P)	Reference
Theoretical	Exploits foods of a single food value (S)	Exploits foods of different values (G)	Stenseth and Hansson, 1979
Theoretical	Restricted to single food taxon ("strict monophagy")	Undefined	Levins and MacArthur, 1969
Butterflies	Restricted to one plant genus (M)	Feeds on plants in >1 family (G)	Slansky, 1976
Butterflies, moths	Restricted to one plant family (S)	Feeds on plants in >1 family (P)	Scriber, 1973, Futuyma, 1976
Insects	Restricted to ferns	Feeds on ferns and flowering plants	Hendrix, 1980
Birds	Concentrates on one or a few categories of food (S)	Utilizes several categories with considerable frequency (G)	Morse, 1971

serve different purposes. Analyses based on the number of higher taxa utilized allow inferences about the extent to which a parasite, grazer, or predator is tracking a host or prey taxon phylogenetically. Analyses based on the number of species eaten allow inferences about the evolutionary dependency of a parasite, grazer, or predator on a particular host or prey.

The second problem in studying patterns of specialization is taxonomic. Several sibling species with narrow host ranges have often been classified as a single species with a much wider host range. Askew (1971) notes that when a parasite thought to feed on many hosts is studied in detail, it is often found to be an aggregate of sibling species each with a much narrower range of acceptable hosts. Fortunately, this is simply a methodological problem that is becoming much less common as interaction increases between ecologists and taxonomists.

The third problem is conceptual. Not all species within a taxon generally considered as parasites, grazers, or predators may actually fit the criteria for one of those designations and, therefore, should not be considered.

together in analyses of specialization. Parasites especially must be considered independently of grazers and predators. How parasites, grazers, and predators specialize must differ because the range of potential evolutionary options differs among them. At the evolutionary moment when a grazer can complete its development on a single host, or when a parasite cannot complete a developmental stage on a single host, or when a parasite kills its host immediately, the bases of selection change. The parasite-turned-grazer or predator can no longer acclimate to its host early in development and respond slowly to any phenological changes that occur in the host's defenses; it must now be able to respond quickly to the array of defenses of several hosts. The grazer or predator-turned-parasite must be able to cope with induced defenses in its host (e.g. Haukioja and Niemelä, 1979; Ryan, 1979). The following sections discuss differences in the evolutionary responses of parasites, grazers, and predators to host and prey defenses.

Parasites

Specialization on single host species or on a few closely related host species is common to many parasite taxa. For example, among North American oecophorid moths in the genera *Depressaria* and *Agonopterix* whose larvae feed on flowers, immature seeds, leaves, or flower stalks, most species feed on only one or a few host species (Table 2.2), although some are capable of surviving on more host species than they feed on in nature (Thompson, unpublished data).

In the most thorough community-wide study of parasite specificity in a tropical area to date, Janzen (1980b) found that a total of 110 species of beetles feed as larvae in the seeds of dicotyledonous plants in the deciduous forests and riparian evergreen vegetation in the lowlands of northwestern Costa Rica. These beetle species are distributed among only 100 of the 975 dicotyledonous plant species and the vast majority (83 spp.) of the beetle species feed on the seeds of only one plant. Among the remaining 27 beetle species, 14 attack two plant species, nine are on three, two on four, one on six, and one on eight plant species. In his book *Evolutionary Biology of Parasites*, Price (1980) provides many other examples of similar patterns of extreme host specialization by parasites.

Specialization on one or a few host species is expected in many parasites because the parasite–host relationship is so intimate. The problem of host specialization in parasites, then, is not to explain why so many species are highly host specific but rather why other species seem not to specialize as consistently on single host species. Parasitic species that are not restricted to one host or a few hosts that share traits important for survival of the

Table 2.2.

Number of Native Plant Species, Genera, and Families Known To Be Fed upon by Larvae in Species of *Depressaria* and *Agonopterix* (Lepidoptera:Oecophoridae) Native to North America[a]

Number Host Plant Species, Genera or Families Known for Each Moth Species	Number *Agonopterix* Species			Number *Depressaria* Species		
	Plant Species	Plant Genera	Plant Families	Plant Species	Plant Genera	Plant Families
1	10	14	16	14	15	18
2	3	3	4	1	1	
3	5	2	1	1	1	
4	1	0		2	1	
5	0	1				
6	0	1				
7	0					
8	1					
12	1					

Source. Data for *Agonopterix* compiled from Hodges (1974); data for *Depressaria* compiled from Clarke (1952 and references cited therein), Hodges (1974) and Thompson (unpublished data).
[a] Moth species whose host plants are known only to genus are excluded.

parasite often exhibit two other patterns of specialization: regional host specialization or specialization on long-lived individual hosts.

Regional Host Specialization. It was first realized over 100 years ago that in species with impressively long lists of recorded hosts, different populations of a parasite often specialize on different host species (Walsh, 1864; Brues, 1924). There are many examples among insects that feed on plants (Knerer and Atwood, 1973; Gilbert and Singer, 1975; Cates, 1980; Fox and Morrow, 1981), and among parasites of animals (Price, 1980). The evolutionary bases for specialization on different hosts in different regions in some instances relate to local or regional variation in the host. For example, the pattern of lycaenid butterfly attack on lupine species depends upon flowering time, pubescence, and alkaloid concentration among the available host plants (Breedlove and Ehrlich, 1972; Dolinger et al., 1973).

Regional differences in the hosts attacked by parasites may also be based on nonplant factors that influence survival on a host such as competition or rates of predation or parasitism on the herbivore (e.g. L. P.

Brower, 1958). Gilbert (1979) suggests the phrase "ecological monophagy" for restriction of parasitic herbivores to one host species even though other potential hosts are available because of ecological, rather than chemical or nutritional, differences among the hosts. He contrasts this phrase with coevolved monophagy for restriction to certain host species through digestive specialization to host chemistry. Although Gilbert's definitions were based on host chemistry, plant defenses other than chemistry could certainly be incorporated into them so that the basic distinction between the two types of monophagy are host versus nonhost influences.

Singer's (1971, 1972) excellent study of oviposition patterns in six populations of the nymphaline butterfly *Euphydryas editha* serves especially to caution against conclusions on parasite–host interactions based on analyses of single populations. In Singer's study the distribution of egg masses on host plants differs dramatically between populations ranging, for example, from 98% of the ovipositions on *Plantago* at the JR (Jasper Ridge) site to 0% on *Plantago* at the IF (Indian Flat) site (Table 2.3). Singer suggests that the predominance of ovipositions on *Plantago* at the JR site resulted from a combination of visual and phenological factors. Ovipositing females choose green areas. *Plantago* is green whereas *Orthocarpus* is magenta from above because it flowers during the butterfly's flight season. Also, alighting females fly to the bases of plants to oviposit. During the insect's flight season, the base of *Plantago* is leafy whereas that of *Orthocarpus* is leafless.

At site EW (Edgewood Road) the pattern is reversed: females oviposit mostly onto *Orthocarpus* rather than onto *Plantago*. Singer suggests that

Table 2.3.
Distribution of *Euphydryas editha* Egg Masses on Host Plants at Six Different Study Sites[a]

Population	Number of Egg Masses					
	Plantago	*Orthocarpus*	*Pedicularis*	*Collinsia*	*Castilleja*	*Penstamon*
JR	40	1	0	—	—	—
EW	3	15	—	—	—	—
DP	—	—	61	4	0	—
IF	0	0	—	25	0	0
MC	—	—	0	12	0	0
GL	—	—	—	—	14	0

Source. Adapted from Singer 1971.
[a]All host plants except *Plantago* are in the family *Scrophulariaceae*. All populations were in California. A dash indicates that the host plant is not present at site.

this pattern results because *Orthocarpus* is 3 – 10 times denser at EW than at JR and because *Plantago* becomes senescent 1 – 2 weeks earlier than *Orthocarpus* at the EW site, making *Plantago* less attractive for oviposition.

Pedicularis is the major host plant at the DP (Del Puerto Canyon) site, but it is not oviposited upon usually at the two other sites where it occurs. This plant is not utilized at the JR site seemingly because of high predation pressure on egg masses on this species at this site selecting against females choosing this host. This same selection pressure works against females ovipositing on *Collinsia* at the DP site. Similar interactions of plant morphology, phenology, predation pressure, and also plant chemistry affect the variations in choice in the other populations. This is one of the few studies that has quantified differences in host selection between populations and made inroads into the ecological basis of the differences.

The implications of this study for analyses of parasite – host coevolution are two-fold. First, parasitic species are often restricted to fewer hosts at the population level than the overall lists of hosts for a species suggest [see Fox and Morrow (1981) for a thorough recent review]. Within a local area the interactions between a parasite and a host may be very specific and coevolved, whereas in another area where both species occur the interaction may be uncommon and subject to little or no coevolution. Second, even within local areas where the interaction between a host and parasite species is common, evolutionary changes in the interaction do not result only from direct coevolution between the parasite and the host. A host may become unacceptable relative to other hosts in a community for a variety of reasons, including differential predation, parasitism, or competition on different hosts (Smiley, 1978) or even, as in the case of the lycaenid butterfly *Ogyris amaryllis*, because some plant species are more likely to harbor ants mutualistic with the butterfly than other plant species (Atsatt, 1981a). In fact, in the case of *Ogyris*, ovipositing females select the nutritionally inferior host *Amyema maidenii* with ants over the nutritionally superior *A. preisii* without ants. Therefore, over evolutionary time host preferences in parasite populations may vary as selection pressures change, some changes relating to parasite – host coevolution and others not. What the ecologist observes as a one-to-one species coevolved unit with a few ovipositional mistakes onto other hosts during a study lasting a few years may be just one point in a cycle that routinely involves musical chairs among a group of potential host plants. That such cyclic shifts in patterns of host utilization may occur is indicated by results of studies showing variation for host preferences among individuals in the same parasite population (Tabashnik et al., 1981; Wiklund, 1981).

Specialization on Long-Lived Hosts. Parasites that feed on relatively long-lived hosts can specialize in a manner different from parasites on relatively short-lived hosts. Some of these differences may decrease the likelihood of coevolution in parasites associated with long-lived hosts, whereas other differences indicate differential rates of coadaptation in a parasite and its host.

Parasites of relatively long-lived hosts have available the evolutionary option of specializing for many generations on a single host individual or on a few closely related host individuals rather than leaving the host after one or a few generations. In the extreme, the effective breeding population for many generations may be the individuals within the body of a large mammal or on a long-lived tree. The parasite deme can potentially adapt to the individual host or a group of genetically related hosts such as a troop of primates or a patch of sibling trees. As a result, adaptation in the parasite population on an individual host can change over time through recombination and selection among the parasites on the host, whereas the host's genotype remains the same [unless somatic mutations occur (cf., Whitham and Slobodchikoff, 1981)]. A study of the interactions of a host and a parasite soon after colonization may produce different results than a study of the same host individual and parasite deme 10 years later.

The interaction between ponderosa pine *(Pinus ponderosa)* and the black pineleaf scale *(Nuculaspis californica)* seems to develop exactly in this way. Edmunds and Alstad (1978) demonstrate that individual trees vary in the defensive phenotypes that they present to scale insects and that these scale insects are divided into semi-isolated demes that track individual pines. Scales show significant differences in their ability to survive on a range of receptor trees during experimental transplants. Furthermore, scale survival is lower in intertree transplants than in intratree transplants. The frequency of males produced within each deme is low and mating is localized within the tree; upon emergence an adult male immediately walks along the needle and copulates with nearby females (Alstad et al., 1980). Therefore, recombination in pines produces a variety of defensive phenotypes and minimizes the chances that a pine's offspring will be attacked by the same scale deme as the parent. Adaptation in the scale insects must be a compromise between tracking an individual host and colonizing new trees. These patterns of specialization and adaptation on individual hosts should be common in parasites able to reproduce for many generations on the same host.

Another evolutionary consequence of feeding as a parasite on long-lived hosts is that under some conditions the potential for further coevolution may be minimal. Parasites of long-lived host species are presented with a broader spectrum of host ages from which to "choose" hosts than parasites on short-lived hosts. A consequence for parasites of choosing among host ages is that parasites attacking older host individuals in iteroparous species are attacking host stages that may not be effectively subject to selection for defenses. This potential safety from defense follows from both Medawar's (1957) and Williams' (1957) logic on the evolution of senescence. Following Medawar's logic, any deleterious mutation that is first expressed after reproduction (including parental care) cannot be selected against since the individual's genetic contribution to the next generation is already complete. This lack of selection against deleterious mutations would include mutations that lowered the efficacy of defense mechanisms. Looked at another way, parasites that attack or produce their most deleterious effects on late or postreproductive individuals could potentially adapt to hosts without coevolutionary repercussions. If defense mechanisms are less effective in older individuals of some host species, then selection for host specificity in the parasite could be reduced and replaced with specificity for particular stages of a variety of hosts.

The bases of host selection on relatively long-lived hosts, then, may be influenced by the population structure (i.e. both deme area and age structure) of the parasite and of the host, and together these may determine if coevolution is likely between the species and the rates of adaptation of the host and the parasite.

Grazers and Predators

Grazers and predators share two potential traits that separate them from parasites in the evolution of patterns of specialization: the potential to become channeled genetically over evolutionary time into requiring a mixed diet, and the potential for individuals to learn which hosts or prey to eat and which to avoid. By definition, neither of these traits influences the patterns of specialization in parasites; hence the evolutionary options available for specialization in grazers and predators differ from the options available to parasites. A requirement of a mixed diet and learning can decrease the extent to which particular pairs of species of grazers and hosts or predators and prey change through coevolution as compared to interactions between parasites and hosts.

Selection for a Mixed Diet. Grazers and predators can be channeled evolutionarily into requiring a mixed diet through selection to maximize the balance of nutrients in the diet, through selection to minimize the effects of toxic or deleterious physical traits of hosts or prey, or through a combination of these. These are problems faced also by parasites; but parasites have evolved to cope with the problems in the context of the advantages and disadvantages that are part of developing on a single host and being dependent upon the fate of the host. Grazers and predators, however, must find several to many hosts or prey and this automatically makes selection for a mixed diet an evolutionary possibility. The time spent searching for the next host or prey individual of the same species as the previous meal may not justify energetically or nutritionally skipping over other potential species. This argument alone suggests selection for the flexibility to feed on several host or prey species, not necessarily selection specifically for a mixed diet. In addition, however, selection may not act effectively to solve nutritional or toxic problems associated with feeding only on one host or prey species if several hosts that may serve as complementary foods are available regularly to the grazer or predator. Under these circumstances, together with the energetic disadvantages of skipping potential host species, selection may act specifically to favor a mixed diet.

Most theoretical models of optimal foraging do not apply to the many organisms that require a mixed diet. Foraging models rely generally on a constant ranking of feeding preferences (reviews in Schoener, 1971; Pyke et al., 1977) or on a ranking based on relative abundance of alternate hosts or prey and switching from one host or prey to another as conditions change (Murdoch and Oaten, 1975). Fewer verbal, graphical, or mathematical models have considered selection specifically for a mixed diet in individual grazers or predators (Freeland and Janzen, 1974; Westoby, 1974, 1978; Pulliam, 1975; Kitting, 1980; Rapport, 1980). These models differ from models based on fixed preferences or switching either by adding a minimum requirement for the nutrients in certain foods, selection for an optimal mix of nutrients (Westoby, 1974, 1978; Pulliam, 1975; Rapport, 1980) or avoidance of foods that have deleterious effects when eaten in large amounts (Freeland and Janzen, 1974; Kitting, 1980).

Some grazers and predators of plants have been shown to survive better on a mixed diet and to maintain a mixture of foods in their diet when given a choice. Small grazers such as grasshoppers often mix two or three host or prey species (Table 2.4). The grasshopper *Melanoplus sanguinipes* has a probability of survival of 24% when fed alfalfa *(Medicago sativa)* alone, and 0% when fed *Cynodon dactylon* alone; on a mixed diet of these plants, survival is 67% (Barnes, 1965). Survival rates and fecundity of the grasshopper *Euthystia brachyptera* are higher when individuals are fed three grass species rather than one (Kaufman, 1965). Similarly, the herbivorous parrotfish *Sparisoma radians* (Lobel and Ogden, 1981) and the sea urchin *Lytechinus variegatus* (Lowe and Lawrence, 1976) fare better on a mixed diet than on a specific diet.

Table 2.4.

Mean Number of Plant Species ($\pm$S.E.) in the Gut of Dissected Individual Grasshoppers and the Total Number of Plant Species[a]

Species	N	Number of Plant Species/Grasshopper	Total Number of Plant Species
Ageneotettix deorum	250	1.9 ± 0.3	12
Arphia conspersa	191	1.6 ± 0.3	11
Arphia pseudonietana	151	1.5 ± 0.3	9
Melanoplus angustipennis	227	3.0 ± 0.4	38
Melanoplus foedus	151	2.7 ± 0.5	34
Trachyrhachys kiowa	200	1.2 ± 0.3	6
Xanthippus corallipes	226	1.9 ± 0.3	14

Source. Data from Ueckert and Hansen, 1971; the seven (of 14) species with largest sample sizes listed here.

[a]Recorded from all examined individuals of each grasshopper species from northeastern Colorado.

N = Number of grasshoppers dissected.

The plate limpet *(Acmaea scutum)* shows a distinct preference in California for a mixed diet of two algal species, consisting of about 60% *Petrocelis middendorffi* and 40% *Hildenbrandia occidentalis* (Kitting, 1980). These percentages remain similar over a wide range of relative availabilities of the two algal species, and no significant differences in the percentage of algal species in the diet occur between individuals. Kitting's (1980) experiments suggest that the mixed diet is maintained partly through avoidance of *Hildenbrandia* as a high percentage of the total diet, possibly because of the excessive tooth wear that results from feeding on this tough alga. It is less clear, however, why *Acmaea* continues to include *Hildenbrandia* in the diet at sites where *Petrocelis* is relatively more abundant. At such sites *Hildenbrandia* is eaten about twice as often as expected by chance.

Some other small herbivores share traits of both parasites and grazers. Although selection in these species does not seem to act to require them to develop a mixed diet, selection seems to act on these herbivores to maintain their ability to feed on several species. Female *Pieris* butterflies oviposit on nearly all cruciferous plants near the Rocky Mountain Biological Laboratory in Colorado (Chew, 1977). Oviposition preferences onto native plants tend to correspond to the suitability of the various crucifer species for larval growth, but individual females probably lay eggs on most of the potential hosts (Chew, 1975, 1977). Chew suggests that females that oviposit only on one of the array of potential crucifer species may not be favored by selection because the relative abundance of the crucifer species is highly variable in space and time.

The feeding habits of the larvae, however, are also critical. *Pieris* eggs are laid often on rosette plants that are not large enough to support development of the larva through pupation. Therefore, some larvae must find at least one additional plant during development. The necessity of larval movement between host plants may prevent selection for restriction to a single plant species (Chew, 1980). The probability among censused plants in Chew's study areas that the neighboring three crucifers were conspecifics ranged from 23 to 87% for the nonlethal crucifers (Chew, 1977). The selection pressures on these butterflies, then, approximate those on a grazer with a limited home range while retaining some parasitic traits as well. Similarly, the pipevine swallowtail *Battus philenor,* which develops on plants in the Aritolochiaceae, cannot complete development on a single individual of its small hosts in eastern Texas. Larvae must discover at least 25 different plants of *Aristolochia* to complete development (Rausher, 1980). Females oviposit on both species of *Aristolochia* although these plant species differ in morphological traits.

In relatively small organisms such as insects, the grazer–parasite interface has been crossed many times, for many hosts are large relative to these herbivores. Large vertebrate grazers, however, seldom live in environments that would make specialization on one or two plant species a viable feeding habit. Selection for larger body size in many mammalian species effectively channels these species into requiring a mixed diet of several to many host species, and one-to-one coevolution between host and grazer species should not be expected in most large grazers. Large grazers of plants often take only a few bites from any one plant individual before moving on to other plant individuals, which may often be other plant species as well. Gauthier-Pilters and Dagg (1981) note that this kind of behavior is characteristic of camels no matter how rich or poor the vegetation, although they can survive for months in a monospecific pasture if so constrained.

The mantled howler monkey *(Alouatta palliata)* is among the largest New World primates and feeds on a mixed diet of plant species each day and throughout the year. During nine months in the field over a two-year period, Milton (1980) made detailed observations of food choice in two 17-member troops on Barro Colorado Island in Panama. The two troops were very similar in feeding habits. Both troops average seven to eight food species each day (Table 2.5). Although a leaf-food species and a fruit-food species may be the same plant species, the number of separate plant species eaten each day averages at least five. About half of the feeding time each day is spent eating leaves, another 40% of the time on fruits, and 10% on

Table 2.5.

The Variety of Plant Species Eaten by Mantled Howler Monkeys *(Alouatta palliata)* **in Two 17-Member Troops on Barro Colorado Island, Panama**[a]

	Old Forest Troop	Lutz Ravine Troop
Mean number of species eaten daily		
Leaves	5.5	4.8
Fruit	1.6	1.9
Flowers	1.0	0.7
All categories	8.0	7.4
Mean daily proportion turnover of food species[b]		
Leaves	0.65	0.59
Fruit	0.40	0.31
Flowers	0.72	0.71
All categories	0.51	0.51
Number of species eaten during study		
Leaves	59	59
Fruit	25	23
Flowers	13	16
Total[c]	73	73

Source. Compiled from Milton (1980).

[a]Data collected during nine months in the field between 1974 and 1976.

[b]That is, the percentage of food species eaten on one day but not eaten on the next day.

[c]Some plant species fit into more than one food category.

flowers. The leaves are almost all young leaves; 3% or less of the time is spent on mature leaves.

Furthermore, the plant species eaten vary daily. In both troops about half of the food species eaten on one day are not eaten the previous day. During the course of her study, Milton recorded each troop feeding on 73 food species, of which 59 were leaves of different plant species. Although the total number of species eaten was the same for the two troops, the plant species eaten were not identical. Only about half the leaf and fruit species and about one fourth of the flower species eaten by one troop were also eaten by the other troop.

The choice of food in these monkeys involves a clear preference for a mixed diet; but they are not random feeders. The top 10 food species account for 59% (Old Forest) to 73% (Lutz Ravine) of the total feeding time, and the Moraceae and Leguminosae seem to be preferred plant families.

At the other extreme from howler monkeys, some relatively large herbivores are often considered fairly host specific in their choices of plants; but even these species utilize at least several plant species. Koalas

(Phascolarcotes cinereus) apparently cannot subsist on any one species of *Eucalyptus* and in captivity are provided with a variety of *Eucalyptus* species (Collins and Roberts, 1978). The giant panda feeds principally on a variety of species of bamboo (Davis, 1964), but the extent to which it can subsist on a single bamboo species is unknown. It sometimes also includes bulbs of alpine plants and some small mammals in its diet (Brambell, 1976). A few predators of animals, however, are specialized to subsist on single prey species. The Everglades or snail kite *(Rostrahamus sociabilis)* feeds only on snails in the genus *Pomacea* and some populations of this bird feed mostly on only one *Pomacea* species (Snyder and Snyder, 1969), although individual birds may rarely take nonsnail prey (Sykes and Kale, 1974).

Freeland and Janzen (1974) argue for large mammalian grazers that highly host-specific feeding patterns should be expected only when several related toxic foods are available in large amounts year round. Extending the argument to predators of animals, genetically based restriction of diet to one or a few related prey species should occur only when efficient handling of the prey requires a specialized morphology, behavior, or physiology and when the prey are reliably available during the times of year that the predator is active (e.g. Everglades Kite, anteaters).

Grazers and predators of plants seem especially prone to selection for a mixed diet when compared with species that feed on animals. The nutritional needs of animals are not coincident with the complement of amino acids, vitamins, minerals, and other compounds found in plants (e.g. Batzli et al., 1980). Also, the rich array of secondary compounds that grazers and predators of plants must ingest may inhibit specialization on single hosts or prey in organisms that must ingest large quantities or numbers of hosts or prey. Whether nutritional or secondary chemistry is the more important selection pressure in the evolution of patterns of specialization is an old debate in the study of parasitism, grazing, and predatory herbivores (Stahl, 1888; Painter, 1951; Dethier, 1954; Fraenkel, 1959, 1969; Thorsteinson, 1960); the debate continues in various forms and for an increasing range of insect and mammalian taxa of grazers especially (Freeland and Janzen, 1974; Westoby, 1974; papers in Montgomery, 1978; Morrow and Fox, 1980). Secondary compounds such as tannins, of course, can affect the availability of nutrients to the herbivore, so the arguments are not independent. No matter what the relative importance of nutrients and secondary compounds is in the evolution of diets in herbivores, these two factors seem to make grazers and predators of plants more likely to be favored by selection into requiring a mixed diet than grazers and predators of animals.

In general, detailed analyses of the diets of individuals are needed if the selection pressures affecting a particular grazer or predator and a particular host or prey are to be understood. These critical kinds of data include (1) the number of individual hosts eaten by an individual grazer during a feeding bout, (2) the proportions of the different hosts (both different individuals of the same host species and different host species) used by an individual during a feeding period, (3) the extent to which individual grazers differ in their utilization of particular host or prey individuals and species within grazer populations, and (4) the degree to which these patterns change with age of the grazers and with seasons. Few studies incorporate more than one or two of these analyses that are necessary to evaluate the evolution of feeding specialization in grazers and predators. Most studies of the food habits of grazers and predators give data for populations rather than for individuals. Populational data may give very little indication of the way selection acts on individuals to select among hosts or prey.

Learning.   Grazers and predators often have an opportunity to learn about their food either from their parents, from other older conspecifics, or by trial and error. For example, vampire bats, which graze on mammals, vary geographically in the hosts they attack and where on the body they bite their hosts, and Turner (1975) suggests that young bats learn how to attack hosts from their parents. The arctiid wooly-bear caterpillars *Diacrisia virginica* and *Estigmene congrua*, both of which move from plant to plant during development, learn to avoid eating plant species that made them ill on a previous encounter (Dethier, 1980). (Illness is identified in these species as excessive regurgitation, convulsions, or partial paralysis among other symptoms.) The closer a grazer or predator gets in its feeding traits to a parasitic habit, the less important learning becomes and the more important selection of hosts based on genetic programming becomes.

The feeding habits of individual grazers or predators are often a learned subset of the range of potential hosts possible for a species and the study of learning is critical in analyses of the evolution of grazer–host and predator–prey interactions. Two- and three-toed sloth species *(Chloepus hoffmanni* and *Bradypus variegatus*, respectively) seem generalized in the number of plant species they eat, but individual sloths are restricted to relatively few plant species and individuals. Individual sloths have a modal tree that is used more than any other tree plus several other trees that are used regularly. On Barro Colorado Island in Panama each of six radiotagged two-toed sloths were recorded in eight or fewer individual trees and four or fewer tree species during 50% or more of radiolocations (Montgomery and Sundquist, 1975, 1978). Similarly, three-toed sloths tend to use eight or fewer individual trees and eight or fewer tree species for the majority of their time. In addition, the array of individual trees and tree species eaten by sloths differs from individual to individual. The particular array of plants eaten by a three-toed sloth in its two hectare or so home range seems to be determined to a large extent by learning from an individual sloth's mother (Montgomery and Sundquist, 1978). Learning which tree individuals are both nutritious and safe to eat may be particularly critical in species such as sloths, which have very low rates of digestion.

In social species, learning which plants to eat or avoid can also be a property of the group. Among 1699 trees available to mantled howler monkeys *(Alouatta palliata)* in a ten hectare riparian forest in Costa Rica, Glander (1978) finds that the monkeys spend 76% of their time in only 88 trees. Not only do the monkeys choose particular plant species but also particular individuals within species. The knowledge of which hosts to use and which to avoid may be based on experience in the troop. The same opportunities are possible in leaf-cutter ants, whose colonies are long-lived. Leaves for the fungus gardens are chosen preferentially from the array of potential plants in the home range of the colony and change seasonally (Rockwood, 1976). Past experience in the colony could influence the patterns of preference.

Of course, food choice in grazers and predators is determined both by learning and by genetic programming, not by learning alone. The purpose of emphasizing learning here is to note a major way in which grazers and predators differ from parasites in their potential bases for food choice. A strong genetic basis for avoidance of some foods has been suggested or demonstrated in some predator species, of which the best documented case is Arnold's (1977, 1980, 1981a,b) study of geographic variation in food choice in the garter snake *Thamnophis elegans*. Both naive newborn snakes and wild-caught snakes from coastal localities in western North America will attack slugs, whereas naive and wild-caught snakes from inland localities refuse them. The differences in food choice between the coastal and inland populations are ontogenetically stable.

Parasites Compared to Grazers and Predators

The evolution of host specialization in grazers and predators, then, differs fundamentally from that in parasites. Specialization on single host species should be the expected norm in parasites as it seems to be (Price, 1980). Restriction to single host or prey species, however, should be uncommon among grazers and predators: the need to find many hosts or prey generates the evolutionary opportunity for requirement of a mixed diet and/or for learning which hosts or prey to eat or avoid. This implies that the bases for coevolution between particular grazers and hosts or predators and prey are even less likely to be understood than those between parasites and hosts when analyzed as two-species systems. Preference or avoidance of a host or prey species may often be relative to the other food species with which it will be mixed in the diet. Furthermore, a newly evolved defense in a host or prey may be met by the grazers or predators with a learned response rather than a genetic response. Therefore, stepwise coevolution between grazers and hosts and between predators and prey may be much more erratic than between parasites and hosts.

PATTERNS OF DEFENSE

The study of defenses has progressed in the past decade from strict descriptions of defense mechanisms to general hypotheses on patterns in the evolution of defense based partially on the previous descriptions. Of course, some specific aspects of defense theory have long received theoretical treatment together with field experiments designed to test hypotheses on defense, of which the most notable is the study of mimicry (e.g. J. V. Z. Brower, 1958a,b; L. P. Brower et al., 1964; Waldbauer and Sternburg, 1976; Jeffords et al., 1979). For the most part, however, suggestions of general patterns in defense are more recent, and the most general hypotheses are these:

1. The level of defense (i.e. the relative amount of resources devoted to defense) should be higher as follows:
   a. In plants in nutrient-poor rather than in nutrient-rich habitats (Janzen, 1974a).
   b. Among plant parts more critical to the plant at that stage of development than among less critical plant parts (McKey, 1974, 1979; Feeny, 1975, 1976).
2. Generalized defenses that minimize the effect but do not prevent an interaction with an enemy should be more common than defenses that usually prevent the interaction in the following:
   a. Species whose individuals have a high probability of being found by enemies (Feeny, 1975, 1976; Rhoades and Cates, 1976; Rhoades, 1979; Maiorana, 1978).
   b. Species whose individuals are likely to be found by many rather than by a few species of enemies (Fox, 1981).
   c. Plant parts that are available for a long period of time rather than a short period of time (Rhoades and Cates, 1976).
3. Among groups of coexisting species whose individuals have a high probability of being found, defenses should be similar, since species

will converge on the few kinds of defenses that can minimize the effect of the enemy on the individual (e.g. digestibility reducers in plants); among groups of coexisting species whose individuals have a low probability of being found, selection should favor divergence of defenses since this decreases the likelihood for a host or prey species that an enemy that can feed on surrounding species will also be able to feed on it (Feeny, 1975, 1976; Rhoades and Cates, 1976).

This sketchy outline of the general hypotheses of defense is perhaps hopelessly simplified, but it does serve to illustrate that the modes of interaction between organisms are not an explicit part of the general hypotheses. My purpose here, then, is to suggest some ways in which the evolution of defense depends upon whether the enemy is a parasite, grazer, or predator. I have chosen to focus on how organisms differ in their evolutionary responses to inevitable (unavoidable) interactions with parasites, grazers, and predators. The other aspect of defense that would make the analysis more general would be a consideration of how patterns of avoidance of enemies are influenced by the mode of interaction, but I have been able to make little headway with this aspect of the problem so far.

An inevitable interaction is one in which avoidance defenses (e.g. camouflage, protean behavior) are ineffective at preventing attack by an enemy. Selection will continually act on parasites, grazers, and predators to overcome avoidance defenses and increase the probability of successful encounters with hosts or prey. If the probability is high that a host or prey individual will interact with an enemy during its life despite its avoidance mechanisms, then selection may act to favor other lines of defense specifically against these inevitable interactions. In these situations selection can act in two ways: either by decreasing the enemy's efficiency or by changing the outcome of the interaction. These kinds of defense relate to the mode of interaction in that (1) selection to decrease the enemy's efficiency in inevitable interactions is more likely to result from attack by parasites or grazers than by predators; (2) selection to change the outcome of interaction occurs with all three modes of interaction, but this defense is effective against predators only in organisms that can lose parts (tillers, branches, seeds, tails) without dying.

Decreasing the Enemy's Efficiency.   The most common way by which plants reduce the efficiency with which herbivores can utilize plant tissues is perhaps digestibility-reducing chemicals, although thorns and spines also serve this function. Digestibility reducers may not affect all herbivores (Fox and Macauley, 1977; Bernays, 1978), but they certainly affect many herbivores. Digestibility reducers can affect herbivores either by impairing growth, reducing fecundity, or lowering resistance to disease (Price et al., 1980). These effects may be interrelated: reduced fecundity may be a consequence of slower growth rates which lead to smaller adults in some insect species. Slower growth rates on low-digestibility diets, however, do not always lead to smaller adult size because herbivores may compensate by eating more plant tissue during development. This compensation produces what Bouton (in Price et al., 1980) calls the paradox of digestibility reducers. If a herbivore responds to digestibility reducers by eating more plant tissues in order to survive and grow, then the presence of these compounds in plants could actually increase the level of damage caused by the herbivore.

This increase in consumption can result in one of two ways. The herbivore can eat more food per unit time and maintain a fast growth rate. Although this is possible for insects feeding on some plants (e.g. Scriber, 1979), it does not seem to be possible for all parasitic herbivores (Scriber, 1978). Scriber (1977, 1979) notes that in plants and plant parts with low water content, such as in the leaves of trees, increasing consumption rates to compensate for digestibility reducers would only aggravate the problem of low water availability for larval metabolism. Therefore, larvae may be forced to simply feed over a longer total length of time in their lives to reach the same adult size as when the digestibility reducers were absent. With either response the damage to the plant is increased by having the digestibility-reducing compound, but the second response—longer total development time—suggests a way in which selection can favor these compounds.

The paradox of digestibility reducers is resolvable when three rather than two trophic levels are considered (Bouton, in Price et al., 1980). Implicit in Bouton's resolution of the paradox is the assumption that digestibility reducers are evolved primarily in response to parasitic herbivores rather than broad-ranging grazers or predators. Prolonging development time in a parasitic herbivore may increase the probability that the herbivore will be killed by predators or parasites during early immature stages (Feeny, 1975; Price et al., 1980). Most of the food eaten by immature insects is consumed in the last two instars (Scriber and Slansky, 1981). Hence, prolonging development time in the early instars may reduce damage to the plant if the early instars are killed by enemies before they reach the later instars.

The only direct tests of how digestibility reducers in plants can influence rates of predation on herbivores and damage to plants have been those of Bouton (Bouton, unpublished; summarized in Price et al., 1980). Two varieties of soybean were chosen for the experiments—Harosoy Normal and PI 80837. Previous studies had indicated that PI 80837 impaired growth rates and lowered feeding efficiencies in the Mexican bean beetle *Epilachna varivestis* in a manner parallel to that of winter moth larvae on diets with or without tannin. Hence, although the actual cause of impaired growth was unknown in the experiments, the effect was that generally attributed to digestibility reducers.

Bouton introduced identical numbers of Mexican bean beetle larvae into cages in the field with one or both soybean varieties. Into some of the cages, he also introduced a predatory pentatomid bug, *Podisus maculiventris*. Bean beetle larvae fed on the high-tannin-analog variety consumed on average 1.26 times that of larvae fed on the normal variety of soybean, but survival in these larvae was somewhat lower (Table 2.6). Even with lower probability of survival on the high-tannin-analog variety, however, the relative total amount of plant tissue eaten was 1.11 times higher in these cages than in the cages with the normal soybean variety. The larger amount of food consumed by each surviving larva on PI 80837 more than made up for the lower probability of survival on these plants. In the absence of predators, the presence of digestibility reducers in the plants led to higher levels of damage to the plants. When predators were added, the situation was reversed. Survivorship of bean beetle larvae on the high-tannin-analog variety dropped from 0.88 to 0.23 when compared with those on control plants, and the relative total amount of plant tissue eaten dropped to 0.29 times that of normal plants. This elegant experiment demonstrated convincingly that digestibility reducers in plants can be an effective defense against parasitic herbivores when considered in the context of interactions among three trophic levels.

Using the logic of these experiments, it would seem that selection for

Table 2.6.
Effect of Predators on Efficacy of Digestibility Reducers as Defensive Compounds in Plants[a]

	Soybean Variety	Relative Amount Eaten/Larva	Relative Larval Survival	Relative Total Amount Eaten
No Predators	Harosoy Normal	1.00	1.00	1.00
	PI80837	1.26	0.88	1.11
Predators	Harosoy Normal	1.00	1.00	1.00
	PI80837	1.26	0.23	0.29

Source. Data from Price et al. (1980).
[a]Soybean variety PI80837 is a variety that produces effects on Mexican bean beetle larvae similar to those of tannins on winter moth larvae. Harosoy Normal is the control variety. The predator is the pentatomid bug *Podiscus maculiventris*.

digestibility reducers would be most effective when prolonged development in the herbivores leads especially to increased predation or parasitism rates on the early stages of herbivore development. Predators or parasites attacking prepupae or pupae of phytophagous insects would be ineffective in selecting for digestibility reducers in plants.

There are, however, some circumstances in which compounds generally considered as digestibility reducers may function effectively without relying on the third trophic level. In these cases the compounds act to kill the herbivore directly or force the herbivore to feed at less desirable feeding sites, rather than as a digestibility reducer. Resins in leaves may decrease growth rates in parasitic herbivores and act as a digestibility reducer, but production of resins in wood can also directly smother the herbivore or cause it to feed at less desirable sites. Extensive feeding by the red pine scale *Matsucoccus resinosae* (Homoptera:Margarodidae) on *Pinus resinosus* during one year causes secretion of resins from feeding wounds in the next year. McClure (1977) finds that nymphs feeding on these trees require 2–3 weeks longer to mature and produce significantly less eggs than their mothers, who had developed on the trees when resin production onto the twigs was lower. Heavy resin production forces the nymphs to overwinter at an earlier substage of the first instar and on one and two-year-old growth instead of on the preferred three-year-old growth. Overwintering mortality is higher for nymphs in first instar substages and for those on twigs younger than three years old. Hence, in this instance, production of resin causes both a prolonged growth rate and forces the nymphs onto overwintering sites in which mortality is higher.

The other circumstance in which digestibility reducers may be effective without the concomitant action of the herbivore's enemies is in interactions between parasites and long-lived hosts. In parasites that tend to adapt to individual hosts and breed as a deme on the same host individual for many generations, digestibility reducers may slow the growth rate in the parasite enough that the rate of parasite population growth on that host plant is retarded. Therefore, evolution of digestibility reducers in long-lived plants could be favored as a defense that slows down the rate of growth of parasite demes once the parasites are adapted to a host individual. This potential relationship between host longevity, parasite deme structure, and digestibility reducers has not been considered explicitly in previous hypotheses on the evolution of digestibility reducers.

In general, what seems clear is that digestibility reducers in plants are evolutionary responses to herbivores that are associated with individual plants for extended periods of time. Parasitic herbivores fall most readily into this category whereas predators do not fit the criteria at all. Possibly grazers that feed on only a few individual plants, moving among these plants on a regular basis, could also select for quantitative defenses.

Individual grazers that failed to avoid plants with high levels of digestibility reducers in their home range may suffer slower developmental rates, as in some parasitic herbivores. For quantitative defenses to be favored by selection at the level of the individual plant, however, consumption rates on that plant must be lower than on plants without the digestibility reducers. This may be possible in a grazer–plant system if the grazer learns quickly which plants to avoid in its home range. Hence, the learning component in grazing may mitigate the need to consider three trophic levels in the evolution of some digestibility reducers. But parasites are still likely to be the primary selection agents for these defenses, although grazers with very limited home ranges may help to select for the maintenance of these defenses within plant populations. Toxins would seem generally to be a more effective mechanism for plants to repel grazers that are likely to avoid plants through learning.

Change of Outcome or Focus of Interactions.   If quantitative defenses are primarily evolutionary responses to parasites and secondarily to grazers with limited home ranges, does this mean that organisms cannot respond to inevitable interactions with predators and broad-ranging grazers except through avoidance? Of course not, but the adaptations in the host or prey are based on changing the mode of interaction with the enemy. If the interaction has a high probability of occurring during the lifetime of a host or prey individual, selection may act to transform a predatory interaction into essentially a grazing interaction.

For example, the native grasses of the steppe of eastern Washington and Oregon suffered heavy mortality with the introduction of cattle in the late 1800s. Unlike the grasses of the Great Plains that had evolved with massive herds of large grazing mammals, the grasses of the Palouse never evolved traits that allowed them to recover after grazing and trampling (Mack and Thompson, 1982). Hence, the interaction between cattle and native grasses in the Palouse is often essentially a predator–prey interaction in the sense that the grasses can be killed quickly by feeding and trampling of large herbivores, whereas in the Great Plains the interaction has evolved into a grazer–host interaction in the sense that individual grasses can sustain being fed upon and trampled without dying.

Some lizards and salamanders have also evolved traits that effectively transform predator–prey interactions into grazer–host interactions. When attacked by a predator, the tail breaks off from the remainder of the body, leaving the predator with a body part instead of the whole body (Congdon et al., 1974; Maiorana, 1977). This kind of defense certainly costs the prey energy and nutrients, but the individual survives the interaction.

An inevitable interaction may be transformed evolutionarily in ways even more dramatic than the change from predation to grazing. Interactions based on parasitism, grazing, or predation may all evolve through a change in outcome to interactions that are either commensalistic or even mutualistic. Many mutualisms have evolved in this manner and these are considered in detail in Chapter 4.

ECOLOGICAL GENETICS OF COEVOLUTION IN ANTAGONISTS

Mathematical models and general ecological thinking on coevolution are based largely on a gene-for-gene approach. The likelihood of strict gene-for-gene coevolution, however, probably varies between modes of interaction. The gene-for-gene concept was introduced by Flor (1942, 1955) to describe the relationship between resistance in varieties of cultivated flax *(Linum usitatissimum)* and the ability to overcome resistance in the flax-rust fungus *(Melampsora lini)*. Flor found 27 resistance genes distributed over five loci in flax, each of which segregated in a 3:1 ratio of resistant to susceptible plants in tests in the F_2 generation, indicating that the resistance genes are dominant. Virulence genes in the rust—that is, genes for overcoming resistance in the plants—also segregated as a 3:1 ratio in the F_2 generation, but always with the virulence gene as the recessive trait. The genes for virulence in the rust, however, do not seem to be localized on a few loci as in the genes for resistance in the plant. Similar examples of gene-for-gene relationships are now common in the agricultural and phytopathological literature (Flor, 1971; Day, 1974; Burnett, 1975; Gallun et al., 1975). Breeding programs for resistance against parasites have in some cases resulted in a coevolutionary race in which plant breeders attempt continually to produce new cultivated varieties as the resistance of the older varieties is overcome by the parasites.

The examples of gene-for-gene interaction involve largely plants and pathogens, but there is also a well-studied example between a plant and a parasitic insect: wheat and the Hessian fly *Myetiola destructor* (Gallun, 1977). Breeding for Hessian fly resistance in wheat has generated gene-for-gene coevolution between wheat and Hessian fly. Eight genes for resistance to Hessian fly have been identified in wheat, of which seven are dominant and one is recessive (Gallun and Khush, 1980). Four recessive genes that allow the flies to overcome resistance in wheat have appeared in Hessian fly populations. These four recessive genes correspond to four of the dominant genes for resistance in wheat. The wheat variety "Turkey" has none of the genes for resistance and is susceptible to all Hessian fly biotypes (Table 2.7). Similarly, the Hessian fly biotype "GP" has none of the known virulence genes to overcome any of the resistance

Table 2.7.
Distribution of Resistance Genes in Some Wheat Varieties and Virulence Genes in Hessian Fly Biotypes, and Their Effects on the Interactions Between Wheat and Hessian Flies[a]

Wheat Variety (Resistance Gene)	Hessian Fly Biotype (Virulence Gene)							
	GP (none)	A (s)	B (s, m)	C (s, k)	D (s, m, k)	E (m)	F (k)	G (m, k)
Turkey (none)	+	+	+	+	+	+	+	+
Seneca (H_7, H_8)	−	+	+	+	+	−	−	−
Monon (H_3)	−	−	+	−	+	+	−	+
Knox 62 (H_6)	−	−	−	+	+	−	+	+
Abe (H_5)	−	−	−	−	−	−	−	−

Source. Adapted from Gallun and Khush, 1980.
[a]Plus indicates that the wheat variety is susceptible to the Hessian fly biotype; minus indicates that it is resistant.

genes in wheat. Hence it can attack successfully "Turkey" but not "Seneca," "Monon," "Knox," or "Abe," each of which has a different dominant gene for resistance. Hessian fly biotype A has a homozygous recessive gene s which allows it to overcome resistance in "Seneca"; biotypes E and F also have virulence genes in the homozygous recessive state allowing them to each attack one of the resistant varieties of wheat. Biotypes B, C, and G have two virulence genes in the recessive state and these can each attack two of the resistant varieties. Biotype D has three virulence genes, corresponding to the resistance genes in "Seneca," "Monon," and "Knox" and, therefore, can attack all three of these varieties as well as "Turkey." No Hessian fly biotype has yet developed a virulence gene in the recessive state to overcome the resistant gene in the variety "Abe" (Gallun and Khush, 1980). The similarity in the genetics of this insect–plant interaction and the pathogen–plant interaction also illustrates the utility of considering coevolution between species based on modes of interaction rather than strictly along the taxonomic lines considered within parasitology and phytopathology journals and texts.

The gene-for-gene concept, also known as the matching gene theory (Gallun and Khush, 1980), is a very specific form of coevolution. Person et al. (1962) define a gene-for-gene relationship as one in which:

"the presence of a gene in one population is contingent on the continued presence of a gene in another population, and where the interaction between the two genes leads to a single phenotypic expression by which the presence or absence of the relevant gene in either organism may be recognized."

In most examples of gene-for-gene interaction, (1) the genes for resistance in the plants are dominant and are often localized on a few multiallelic loci, whereas (2) the genes for virulence in the parasite are recessive and nonallelic (Person, 1967; Gallun and Khush, 1980; Sidhu and Webster, 1981). Also most of the examples involve plants with multiple alleles for resistance. Although the known examples involve cultivated plants and their parasites, the multiple genes for resistance and virulence may be maintained within natural populations of hosts and parasites by genetic polymorphisms. These polymorphisms could be maintained through heterozygote advantage in hosts with dominant alleles for resistance (e.g. Mode, 1958), frequency-dependent selection (e.g. Person, 1967; Clarke, 1976), and/or density-dependent selection (e.g. Clarke, 1976).

The suggestion that selection in gene-for-gene interactions may act to maintain genetic polymorphisms in both the host and parasite is common in mathematical models of coevolution (Roughgarden, 1979; Slatkin and Maynard Smith, 1979). In contrast, ecological studies of coevolution have focused mostly on an approach where the underlying assumption is stepwise coevolution through directional selection. Only long-term studies of interactions with species in which the genetic bases of the interaction are known can tell us if gene-for-gene interactions favor genetic polymorphisms in hosts and parasites. Short-term studies demonstrating directional selection in a host caused by a parasite may be only one point in time in a frequency-dependent interaction. It is conceivable that coevolution of hosts and parasites involves rare additions of new alleles that allow for directional selection, and that at other times selection occurs as a frequency-dependent or related process in which a battery of defenses and counterdefenses wax and wane in their effectiveness within populations. Clarke (1976) even suggests that interactions between parasites and hosts may be a dominant force in maintaining protein polymorphism within populations.

Genetic polymorphisms are also thought to be maintained at least partially by predators in some prey taxa such as *Cepaea* snails (Jones et al., 1977). The process here, however, involves genetic change in the prey and learning in the predator, rather than gene-for-gene coevolution. In general, gene-for-gene coevolution is probably much less likely in predator–prey and grazer–host interactions than in parasite–host interactions, because predators and grazers can respond to new defenses through learning. Avoidance of aposematic animals by birds (e.g. Jeffords et al., 1979) is often a learned response (e.g. J. V. Z. Brower, 1958a,b), although it can become genetic (Smith, 1977). Large herbivores may nibble on a wide variety of plants and learn which are safe to eat within their home ranges

(Freeland and Janzen, 1974). Therefore, grazers and predators may evolve through their interactions with hosts and prey but it may seldom be in the specific form of the gene-for-gene concept.

All this is not to say that parasites cannot learn. Ovipositing parasitoids can learn apparently where to find hosts, or at least they can increase their efficiency over time at finding hosts (Alloway, 1972; Taylor, 1974; Cornell and Pimentel, 1978). Nevertheless, the choice of which kinds of hosts its larvae should eat or avoid is a genetic choice. There is no immediate feedback to the ovipositing parasitoid, only selection through larvae that live or die.

CONCLUSIONS

This chapter has suggested the following patterns in how specialization, defense, and the ecological genetics of coevolution vary according to whether the enemy in an interaction feeds as a parasite, grazer, or predator:

1. Most parasites feed on only one or a few host species although in some parasite species different populations feed on different host species.

2. Parasites of long-lived hosts can adapt to individual hosts, and some parasites attacking late or post-reproductive hosts may be able to adapt to hosts without any possibility of favoring new host defenses that would be likely in parasites attacking younger hosts.

3. Selection can act on grazers and predators to require a mixed diet.

4. Specialization on certain hosts or prey in grazers and predators can result from learning and can differ between individuals and populations through learning rather than through genetic differences.

5. Some species at the border between parasites and grazers (e.g. caterpillars that eat several plant individuals during development) may be favored by selection to maintain the ability to feed on a broader range of plant species than related parasitic species that can complete development on a single host.

6. Hosts can respond evolutionarily to inevitable interactions with parasites or grazers with defenses that reduce the efficiency with which the enemy can harvest resources. Digestibility reducers are the most common chemical defense of this type in plants. The evolution of these compounds depends upon one or more of the following:

    a. The enemy is a parasite and is subject to high rates of predation in the larval (nymphal) stages.

    b. The parasite deme adapts to individual hosts and the digestibility reducer decreases the rate of increase of the parasite deme on that host individual.

    c. The enemy is a grazer with a limited home range.

7. Some prey can respond evolutionarily to inevitable interactions with predators through defensive traits that effectively change the outcome of the interaction into a grazer–host interaction by allowing detachment of some body parts such that death does not result from the interaction (e.g. some rhizomatous grasses, tails of some salamanders).

8. Gene-for-gene coevolution is likely to be more common between parasites and hosts than between predators and prey or between grazers and hosts.

CHAPTER

3

COMPETITION AND COEVOLUTION

The underlying importance of interspecific competition in community structure and in the evolution of life histories has been a pervasive theme in the development of ecology. The belief that sympatric competitors coevolve has also been part of the general theme. Nonetheless, interspecific competition and coevolution have been inferred more often from observation than from experiment. This is unfortunate because competition is unlike other kinds of antagonistic interaction, and experimental studies of coevolving competitors can provide interesting contrasts to coevolving antagonists at different trophic levels.

Competition differs fundamentally in one critical respect from all other kinds of interspecific interactions that can lead to coevolution: selection does not act on any competing species specifically to increase the likelihood that an individual will encounter a competitor. I think this makes competition generally less likely to lead to long-term coevolution of particular sets of species than any other form of antagonistic or mutualistic interaction. Selection acts on parasites, grazers, and predators to increase their ability to find suitable hosts or prey; selection acts on mutualists to increase their probability of interaction. No comparable selection pressure generally holds competitors together. Once one of the species diverges from a competitor in its use of resources, the interaction ceases (or at least decreases) as a selection pressure on the species and there is probably little selection on the other species to diverge also. Therefore, coevolutionary divergence of competitors is probably much less common than divergence of only one of the competitors (Figure 3.1).

This chapter begins by considering the kinds of mostly nonexperimental studies that have inferred competition and coevolution in natural populations. The remainder of the chapter focuses on the kinds of organisms, communities, and competitive interactions that are most likely to lead to continuing coevolution between two or three species. If we can specify the ecological conditions that are likely to result in coevolution of competitors, then it becomes more likely that we will be able to assess the role of competition relative to other modes of interaction in the evolution of species and in the interaction structure of communities.

Throughout this chapter competition is considered to act in two ways: exploitation and interference (sensu Miller, 1967; Case and Gilpin, 1974). In exploitation competition, species with higher efficiency at exploiting limited resources compete indirectly with species with lower efficiency by lowering the amount of resources available to the poorer competitor. In interference competition, the better competitor excludes directly the poorer competitor from access to the resource. Alternative uses of the terms competition and interference are clarified by Tinnin (1972).

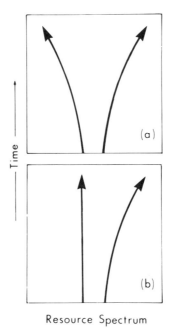

Resource Spectrum

FIGURE 3.1. Possible patterns of evolutionary divergence between two species initially similar in their use along a resource spectrum. (*a*) Coevolutionary divergence. (*b*) Simple evolutionary divergence in only one of the species.

INFERRED COMPETITION

Interspecific competition has proven difficult to demonstrate in natural populations partly because some organisms are difficult to manipulate in controlled experiments. Yet a massive number of studies have inferred competition based on nonexperimental methods, and the sheer bulk of these studies creates the impression that interspecific competition is the most important kind of interaction shaping life histories, populations, and communities.

Nonexperimental studies have inferred competition from either of two kinds of data. The first is negative: failure of a species to utilize a particular resource or habitat is assumed to result from either specific or diffuse competition with other species. These studies take a restricted form of the White Knight's dilemma in Lewis Carroll's *Through the Looking Glass:*

> "But you've got a bee-hive—or something like one—fastened to the saddle," said Alice.
>
> "Yes, it's a very good bee-hive," the Knight said in a discontented tone, "one of the best kind. But not a single bee has come near it yet. And the other thing is a mouse trap. I suppose the mice keep the bees out—or the bees keep the mice out, I don't know which."

The White Knight's dilemma, of course, is an absurdly extreme form of inferred competition from failure to use a resource, since neither species in this case uses the "resources" it is expected to use and it is unlikely that these are resources for which they could compete. (It is important to note here that the resource for the mouse is assumed to be the bait in the mouse trap rather than the trap itself.)

In actual studies competition is often inferred based on negative data because a species or group of species that uses a certain resource in one part of its geographic range either does not use that resource in another part of its range or is uncommon when in sympatry with a purported competitor. Some studies of this type infer competition based on detailed knowledge of the natural history of the taxa (e.g. Benson, 1978; Kohn, 1978), whereas in other studies competition seems to be simply the assumed answer for failure to utilize resources. As another variation on this theme of how failure to use a resource relates to competition, other studies assume that any differences observed in resource use between sympatric closely related species are caused by competition. Many studies of what is often called resource partitioning are of this type, including many of the studies reviewed by Schoener (1974). The important point is that all of these studies interpret failure to use a particular resource as evidence for competition now or, as Connell (1980) puts it, because of the ghost of competition past.

In contrast, the other kind of nonexperimental study to infer competition uses high overlap in resource use between species as an index of the intensity of competition. This type of study is related to those that interpret failure to use a resource as evidence of competition in that species that overlap greatly in resource use are expected to diverge over evolutionary times to lower the intensity of competition; these species are expected eventually to fail to use certain resources. The notion that two species identical in all respects in their use of resources cannot coexist indefinitely is an old idea in ecology dating back to the mathematical treatments of Volterra and Lotka in the 1920s and 1930s and the experiments of Gause (e.g. Gause, 1932). This general idea, however, has often been translated effectively into an expectation that considerable overlap on any resource axis suggests intense competition between species. There is no a priori reason to expect this relationship: high levels of resource overlap can indicate that the resource is not limiting and that the level of competition for the resource is low (Dayton, 1973; Menge, 1979). Lister (1980), for instance, notes that in the five species of *Dendroica* warblers inferred by MacArthur (1958) to segregate spatially in order to avoid competition, six of the 10 possible pairs overlap by 0.5 or greater and such high overlap may indicate abundant rather than scarce resources.

Together, competition inferred from high resource overlap and from failure to use resources cover the whole gamut of possible similarities and differences between species. The ease with which explanations invoking competition can explain any pattern of resource use among species cautions against any inferences concerning competition not based on experiments (Dayton, 1973; Connell, 1975, 1978; Menge and Sutherland, 1976; Wiens, 1977; Hairston, 1980c; Lawton and Strong, 1981).

INFERRED COEVOLUTION

Inferring coevolution among competitors is at least twice as difficult as inferring competition: both (or all) species must be shown to have evolved in response to the interaction. This point is fairly straightforward, yet it is seldom considered critically in studies claiming coevolution of competitors. In fact, rather than coevolution being one possible outcome of a competitive interaction, it is often assumed to be the only or at least the most likely outcome of competition (see also Case, 1979; Connell, 1980).

For example, character displacement has been studied by comparing ratios of morphologies in groups of related species within communities. Hutchinson (1959) finds that sympatric species of closely related birds or mammals tend to have ratios of their trophic apparatus ranging from 1.1 to 1.4, the mean ratio being approximately 1.3. Similar ratios in series of coexisting species have been described subsequently for a variety of organisms (e.g. Schoener, 1965; Price, 1972, 1975a; Pearson and Mury, 1979). The basic conclusion of these studies generally is that there is a limiting similarity in coexisting species, given by their observed morphological ratios. Few studies, however, have compared the results of similarity between coexisting species with the ratios from null communities or guilds constructed by assembling species drawn at random from the pool of species in surrounding communities (e.g. Inger and Colwell, 1977; Connor and Simberloff, 1979; Lawlor, 1980). Strong et al. (1979) analyzed avian communities of the Tres Marias and California Channel Islands and suggest that the actual communities do not differ significantly from the null communities (but see Grant and Abbott, 1980). Such analyses are important in that they help separate random from evolved aspects of community structure. The problem is in how to design properly the null community (Pianka et al., 1979; Lawlor, 1980; Hendrickson, 1981; Strong and Simberloff, 1981) and how to evaluate whether a trait deviates significantly from an expected ratio between species (Roth, 1981; Wiens and Rotenberry, 1981).

Even if a consistent ratio in morphology could be shown to differ significantly from a null community, however, this would not automatically indicate that coevolution between particular species had occurred. Imagine three species along a resource axis so that there is competitive room for two other species to invade between the other pairs of species but not without causing some competition for the resources. If the original three species are relatively strong potential competitors and the invading two species are relatively weak competitors, then the final community of five species may involve only the evolution of the invaders to fit into the community and may not involve any coevolutionary change between any particular pairs of species. Coevolutionary scenarios are also possible (e.g. Roughgarden, 1976; Slatkin, 1980). The point here is not to exclude those scenarios but rather to argue that coevolution of particular sets of species cannot be assumed to result from competition and certainly not from comparisons of ratios of traits among closely related species.

Character displacement has also been studied by comparing morphologies of species pairs in allopatry and sympatry. Grant reviews the evidence for a competitive basis for character displacement in general (Grant, 1972) and for the classical example of *Sitta* nuthatches (Grant, 1975) and concludes that the evidence is weak. He notes that the principal difficulties in demonstrating character displacement resulting from sympatry are in identifying the original and derived populations, identifying the precontact character states, and predicting the character state in the zone of sympatry in the absence of character displacement. The overall problem is that to attribute character displacement to sympatry, an investigator must be able to eliminate other variables that could cause change in characters in those regions; differences in characters in the zone of sympatry may result from coincident changes in physical environmental variables or prey resources such that the changes in sympatry are independent of competition.

Since Grant's review, several studies have attempted to demonstrate character displacement between species pairs. Some studies have suggested character shifts in only one of the interacting species (e.g. Huey et al., 1974), whereas several others have implied specifically a coevolutionary response in the zone of sympatry (e.g. Case, 1979; Dunham et al., 1979). Among the few experimental studies of character displacement as a coevolutionary response to competition is Fenchel's analyses of two species of mud snails of the genus *Hydrobia* that are sympatric over part of their range in northern Jutland. In sympatry *H. ventrosa* has a smaller body size than *H. ulvae* than in allopatry, whereas *H. ulvae* is larger in sympatry than in allopatry (Fenchel, 1975). When the species are allopatric they are approximately the same size. Sympatric populations of these two species also have shorter and more well-defined reproductive periods than do allopatric populations. Two other species of hydrobiid snail, *H. neglecta* and *Potamopyrus jenkinsi*, also coexist in some areas with these species. *Hydrobia neglecta* is rarely allopatric from the other so that evaluation of character displacement is difficult; *P. jenkinsi* selects food of sufficiently different sizes from the *Hydrobia* species so that character displacement in sympatry is unlikely.

In laboratory experiments, Fenchel and Kafoed (1976) demonstrate that exploitation competition does exist between the species among individuals of the same size and the displacement in size would lower competition for food. (The snails are deposit feeders and utilize diatoms.) In the laboratory experiments, interspecific competition and its effect on growth of snails was as intense as intraspecific competition, thereby indicating that interspecific competition could be an important selective force. Also, individuals of the same size ingested the same sized particles in all *Hydrobia* species, and individual growth of snails was correlated with the availability of diatoms of particular sizes. Together these results indicate that species with the same size frequency distribution have nearly complete overlap in sizes of the food they select and they compete for the resources. What is missing from the analysis, of course, is a demonstration that food is limiting in natural populations. Also the extent to which the populations mix from year to year is unknown. The potential for coevolution, however, occurs between these snails.

In general, inferring coevolution from morphologies or behaviors of organisms is a risky task, full of potential pitfalls and alternate interpretations. Few studies have analyzed experimentally traits in species that appear to have coevolved. Coevolved competitors seem to be an elusive lot, but the following four sections attempt to narrow the search to competitors that are most likely to coevolve over a long term.

PROBABILITY OF ENCOUNTER

I think that many studies of competition involve searches for coevolution among the kinds of organisms least likely to be tightly coevolved—highly mobile organisms in species-rich environments. Two potentially competing species are unlikely to coevolve if the probability of encounter in any generation is low. That seems obvious. What seems less obvious is that a low probability of encounter between any two particular species may be a common trait of species-rich communities. Connell (1980) argues precisely this point, emphasizing that coevolution between competitors is likely primarily in communities composed of few competing species. In such communities any two species have a high likelihood of interacting, whereas in species-rich communities species often interact with a wide array of other potential competitors.

For example, Talbot et al. (1978) constructed artificial reefs of varying structure in the Great Barrier Reef and report that 105 species of coral reef fish colonized these reefs over the years of their study. Even allowing for some degree of habitat specialization among these species, an individual of one species, colonizing a newly opened space in the reef, may encounter any several of dozens of other species as nearest neighbors. Certainly, the probability of encounter between any two particular fish species must be low.

To Connell's argument that coevolution between competitors is most likely in species-poor communities may be added another condition increasing the likelihood that particular competitors will coevolve: species whose individuals have little control over where they become established and cannot move once established are more likely to coevolve than highly mobile species whose individuals can choose where to nest or where to forage. The reasoning here is that organisms that can choose precisely the areas in which they will grow, feed, or reproduce can avoid potential competitors, especially those of better competitive ability. Admittedly, this avoidance may be the result of past competition—but the point is that in highly mobile species the avoidance may well be one-sided rather than a property of both species in a competing pair. The fairly large number of linear competitive hierarchies found between species suggests that this one-sided outcome is often likely (see below).

If species cannot choose precisely where they become established, however, then the probability of interaction between the species remains high. Potential examples include species with wind-dispersed seeds or planktonic larvae. When the probability of encounter is high and essentially inescapable, then increased competitive ability may be favored by selection and continuing coevolution may be the result. Hence the most likely ecological situations in which to search for coevolution between particular sets of competitors would seem to be among (1) relatively sessile organisms (2) with imprecise abilities to choose specific sites in which to live (3) in low-diversity communities.

As in many of the arguments throughout these chapters, however, these conditions do not guarantee coevolution. Rather these conditions serve only to increase the probability of coevolution among competitors from what, in general, seems to be a low probability. Even if some of these conditions prevail, strong continual competitive interactions between any two or more particular species in a community may be unlikely because many local communities are in a continual state of nonequilibrium or at low equilibrium in which interspecific competition is unlikely.

The equilibrium–nonequilibrium dichotomy in community theory has its roots in the old concepts of climax versus successional communities and the relative ease of mathematically modeling communities in which population growth rates approach zero. In communities approaching equilibrium, disturbance rates must be low and populations may approach carrying capacity for the component species. Interspecific competition may be common. Alternatively, it may be uncommon because predators and parasites in these communities keep populations at a relatively low level. Connell (1975) argues that predation appears to be more intense in benign than in harsh physical environments and, therefore, competition should be more often prevented in these environments than in harsher physical environments. And yet it is mostly in relatively benign physical environments that the search for closely coevolved competitors has been made (e.g. Terborgh and Faaborg, 1980).

But is competition actually more likely under harsh physical conditions when predators and parasite populations are kept at low levels? Wiens (1977) argues that environments considered as relatively harsh physically are usually variable in resource levels both within the lifetimes of individuals and over a longer time-scale as well. Under these conditions it may be difficult for populations to closely track changes in these resource levels. Therefore, populations will often be below the levels at which interspecific competition is likely to occur on a regular basis.

The variability in resource levels in many communities and in the population levels of potential competitors is often the result of physical disturbances that are an inherent part of all communities (Levin and Paine, 1974; Connell, 1978; Pickett and Thompson, 1978). Hence a combination of physical disturbances and enemies at higher trophic levels may often act to decrease the probability of encounter between potential competitors, regardless of whether the conditions approach a local equilibrium. Lawton and Strong (1981) argue that a low probability of competition is the general rule in leaf-feeding insects.

This scenario also seems to be applicable to the seastars *Asterias vulgaris* and *A. forbesi* that live sympatrically off the coast of New England. These species overlap to a large extent in all resource dimensions usually measured in studies of interspecific competition: habitat use, feeding time, body size, prey species, and prey size (Menge, 1979). Nonetheless, the species appear to compete for food only rarely. The populations of the two species usually seem to be held below competitive levels by a combination of large scale mortality from periodic storms and disease and the patchy distribution of food resources, especially mussel beds. These results, however, were a contrast to Menge's (1972) results for starfish along the coast of the northwestern United States. In that system the starfish are the top carnivores and periodic storms, disease, and prey patchiness do not act to keep these species below competitive population levels. The differences Menge found between the Atlantic and Pacific coasts in the probability of competition between starfish species cautions against knee-jerk generalization of competition found within a taxon in one community to all communities in which that taxon occurs.

In summary, for coevolution to occur between particular sets of competitors, the probability of encounter between the species must be high. The characteristics of many species and communities may not often favor these conditions.

COMPETITION FOR MUTUALISTS

Species compete for a wide variety of resources, but some resources may be more likely than others to keep potential competitors interacting over evolutionary time. Of these resources, mutualists may be the kind of resource most likely to engender and maintain coevolution among a set of competitors. The kind of mutualist I envision here is one that must visit several to many host individuals during its lifetime as, for example, aphid-tending ants, cleaner fish, and pollinators and seed-dispersal agents of plants. These kinds of mutualists are seldom restricted to a single host species. The logic behind this assertion is developed in detail in Chapter 6, but the major point is that mutualists that must visit several host individuals usually visit several host species for many of the same reasons that grazers and predators (as compared with parasites) utilize at least several host or prey species.

Although such mutualists usually visit several host species, they may visit one host species more often than others [e.g. *Formica* ant species tending aphid species (Addicott, 1978, 1979)]. This quantitative difference may generate competition among the host species for mutualists (Howe and Estabrook, 1977; Addicott, 1978), and may be the basis for continuing coevolution among the host species. Individuals of either competing species that possess traits increasing the probability of being visited by the mutualist will be favored by natural selection. The interaction between the competitors is maintained by the mutualist, which may move freely between the competing species in both ecological and evolutionary time as new traits develop in the host species. Unlike in some other forms of competition, the mutualist can act to hold the competitors together over evolutionary time.

This section focuses on competition for pollinators as an example of how mutualists are a resource that may generate continuing coevolution among small sets of competing species. I chose this type of mutualism for the example because the potential evolutionary outcomes of competition for pollinators are variable and because these mutualisms provide some of the best opportunities for continual coevolution among competitors. The studies I discuss do not actually demonstrate coevolution among plant species competing for pollinators but serve only to suggest how pollinators can potentially provide the conditions for long-term coevolution.

Plants may diverge from other sympatric species in their use of pollinators or timing of flowering for the following reasons: (1) competition for attractions of pollinators or (2) selection to reduce interspecific pollen movement (reproductive isolation). Generally, competition for attraction of pollinators is competition for a limited resource in the sense used in most studies of competition. Selection for reduced interspecific pollen movement could occur even if pollinators were not a limiting resource. It is helpful to separate selection resulting from competition for attraction of the mutualist pollinator from selection to reduce interspecific pollen movement in analyses of plant–pollinator interactions because the evolutionary outcomes may differ.

Competition for a limited number of pollinators can only be overcome evolutionarily by selection for traits that attract more pollinators (unless pollination by animals is abandoned altogether). Such traits include increases in nectar production relative to competitors or divergence in flowering time from competitors. In contrast, selection for reduced interspecific pollen movement need not involve divergence in most floral traits or phenology except in placement of the anthers and stigmas. If plant species differ in where pollen is deposited on the pollinator's body (e.g. Dressler, 1968), then the plants can share the same pollinator; if pollinator visits are not limiting pollination, then the plant species should not be expected to diverge in other traits such as flowering time. Competition for attraction of pollinators and selection for reduced interspecific pollen movement, however, merge in situations in which the pollen of different plant species cannot be distributed differentially over the pollinator's body, and the presence of pollen from different species on the stigma interferes with pollination.

Hence, the expected evolutionary outcome of situations in which the same pollinator visits two or more plant species depends upon the extent to which pollinators are limiting, the effect of interspecific pollen on the individual plant, and the degree to which it is possible to place pollen differentially on parts of the pollinator's body. Competition for pollinator visits is only one of several factors influencing divergence in pollinator use (Inouye, 1978) or in floral phenology (Heinrich, 1975).

Nevertheless, competition for pollinators can be a potent selective force for continuing coevolution between plant species because pollinator species are seldom specific to single plant species. Even divergence in flower types does not necessarily eliminate competition for pollinators; such divergence may reduce interspecific pollen movement but may not affect competition for attraction of pollinators. For example, Heinrich (1975) emphasizes that bumblebee-pollinated plants that flower at the same time differ greatly in floral shape, and this seems to be a result of selection to reduce interspecific movements by foraging bumblebees. Individual bumblebees are highly constant to individual plant species during a foraging trip (Grant, 1950; Heinrich, 1976), at least partly because bees learn how to collect nectar and pollen more efficiently on particular plant species over time (Heinrich, 1979). Therefore, selection should act on bumblebee-pollinated plants to maximize differences from coflowering species so that a bumblebee is less likely to confuse an individual of a plant species with an individual of another plant species. Bumblebee colonies, however, are comprised of many individual workers, and individuals differ in the flower species they visit (Heinrich, 1976). Some plant species may attract a larger number of bumblebee workers than other species, so competition may continue over evolutionary time as plant species seesaw in their ability to attract a larger share of the available bumblebee individuals.

The interspecific competitive interaction is held together effectively by the foraging flexibility of the bumblebees, although not all of the changes over evolutionary time in the ability of species to attract pollinators are caused by interspecific interactions. Some of the changes may be tortuitous outcomes of sexual selection acting within the plant species. Individual plants that can attract more pollinators than conspecifics may be favored because they father more offspring on neighboring plants or because the presence of pollen from several fathers on a stigma allows for active or passive female choice of fathers (Willson and Rathcke, 1974; Janzen, 1977a; Willson, 1979; Bawa, 1980).

Not all interactions between plants and a shared pollinator result in either divergence or no change in most floral traits. Hummingbird-pollinated plants in parts of the western United States seem to have actually converged in most floral traits (Grant and Grant, 1968). Brown and Kodric-Brown (1979) demonstrate that where hummingbird-pollinated plants have converged in floral traits such as shape and color, however, they are different in the orientation of anthers and stigmas. The conditions that select for such convergence in floral shape and color so that the pollinator visits the several species indiscriminately are not readily traceable to one cause.

Brown and Kodric-Brown suggest several characteristics of temperate hummingbirds and hummingbird-pollinated plants that act in concert to select for convergence. First, temperate hummingbirds are migratory and must visit several to many flower species during the course of a year. It is advantageous to a hummingbird-pollinated plant to employ signals similar to other hummingbird-pollinated plants so as to attract any hummingbird that is available locally (Grant, 1966). Second, temperate hummingbirds are generally both intra- and interspecifically territorial. As a result, several species of hummingbird-pollinated plants in a local area may have only one hummingbird available reliably. Third, the position of hummingbirds relative to flowers as they hover is highly predictable, making it possible for plant species to diverge in placement of anthers and stigmas and reduce interspecific pollen movement. Fourth, the extent to which local populations of hummingbird-pollinated plants are limited by pollinator availability or subject to interspecific pollen movement varies greatly. Abundance and species composition of hummingbird-pollinated plants differs locally and seasonally, so populations may be subject to deleterious interspecific interactions only erratically. Finally, rare hummingbird-pollinated species may benefit especially from converging with other more abundant species. Otherwise, the plant species might be avoided by hummingbirds because their low numbers would not provide sufficient energy for the high energy demands of the hummingbird lifestyle. [Schemske (1981) develops a similar argument for convergence of floral traits in two bee-pollinated *Costus* spp. in central Panama.] Brown and Kodric-Brown suggest that this suite of characteristics makes temperate hummingbird-plant interactions different from other pollinator-plant interactions, many of which show differences in many floral traits among coexisting species; tropical hummingbirds are nonmigratory, and those birds exhibit a much wider variety of foraging techniques than their temperate counterparts; bumblebees are not migratory, territorial, or large enough generally to carry different pollen on different parts of the body.

Deciphering which plant species acted as the model among the temperate hummingbird-pollinated plants upon which the other species converged is probably impossible. The plant species studied by Brown and Kodric-Brown range over seven families. That any one species acted as a static model cannot even be assumed. Unlike many kinds of interactions based at least in part on competition, however, the interaction is likely to be held together over evolutionary time, as among bee-pollinated plants, through the behavior of the hummingbirds.

In the preceding paragraphs the general argument for why competition for mutualists is more likely to lead to continuing coevolution than other forms of competition is based on interactions in which the mutualist visits several to many hosts. But why are interactions involving a parasitelike mutualist that spends all or most of its life on a single host less likely to be effective in maintaining long-term coevolution between competitors? The reasons follow from the analogy (and often the homology—see Chapter 4) between the different forms of mutualism and parasitism, grazing, and predation. Mutualists associated intimately with their hosts over a major portion of their lives are likely to become more specific or remain specific to single host species over evolutionary time for many of the same reasons that parasites are highly specific to single host species. Mutualistic mites are dispersed to food sites by their beetle hosts and are carried by the beetles, because the mites kill the fly larvae that compete with the beetle larvae (Springett, 1968). These mites must have life histories finely tuned to those of their beetle hosts. Still other forms of mutualism that involve parasitelike mutualists do not provide any apparent possibility for competition for the mutualists. Examples include gut symbionts of ruminants or fungal mutualists of *Atta* ants. Therefore, although competition for mutualists is a kind of competition likely to lead to continuing coevolution between the competitors, not all kinds of mutualism are equally likely to generate or maintain such competition.

COMPETITIVE HIERARCHIES

The previous two sections suggested two sets of patterns regarding the ecological conditions under which competitors are likely to coevolve continually: (1) potential competitors with weak abilities to choose precisely where they become established in communities with low species richness are more likely to coevolve continually than highly vagile species in species-rich communities; and (2) the limiting resource most likely to generate continual coevolution between competitors is mutualists, especially mutualists that visit at least several host individuals during a lifetime. These final sections on the ecological conditions likely to generate continual coevolution among competitors focus on how the likelihood of coevolution is influenced by the structure of competitive hierarchies.

The question is: what kinds of competitive hierarchies are most likely to be the raw material for coevolution? This question is not independent of those asked in the previous two sections, but allows yet another way to analyze the problem of coevolving competitors. The potential ecological outcomes of interspecific competition span the gamut from predictable exclusion of one species by another in a competitive hierarchy to highly variable outcomes among similar species. Not all of these kinds of competitive interactions, however, may generate continual coevolution between species.

The sorts of data that best begin to address the question are few. The most convincing evidence for interspecific competition is provided by controlled field experiments on organisms not constrained to pens in which one species is removed from or added to a natural community (Reynoldson and Bellamy, 1971; Connell, 1975; Colwell and Fuentes, 1975). Some field experiments have demonstrated competition by removing one species or group of species and observing changes in another species relative to control areas (e.g. Connell, 1961a; Williams and Batzli, 1979; Dhondt and Eyckerman, 1980; Munger and Brown, 1981) or by observing interactions in a community over time, removing one species, then observing changes in the other species in the same community, but with no control (e.g. Davis, 1973). These field experiments, especially those with controls, provide evidence that interspecific competition occurs between some species; but to unravel how, where, and when competition is likely to lead to continuing coevolution, reciprocal field experiments are needed, and these are even fewer in number than controlled experiments in which one species is removed.

In a two-species interaction, reciprocal field experiments on competition require at least three treatments: removal of species A only, removal of species B only, and a control area with both species, preferably all replicated. The addition of resources to a study area as a separate treatment can provide information on the relative abilities of the species to capture resources. Such experiments indicate (1) whether the interaction affects both species negatively, and, if so, (2) the extent to which the effect of the interaction is asymmetric. That is, does species A have a greater effect on species B than species B has on species A? The effect can be either changes in population levels or shifts in use of resources or habitats. These different effects may affect the potential evolutionary outcome of competition in different ways (Thomson, 1980), but the important question here is only how commonly asymmetric effects occur. Whenever possible, the examples in this section are from reciprocal field experiments, but as will become clear immediately, these will be in the minority of studies cited.

Linear competitive hierarchies, in which species A almost always displaces species B when in competition are a common result in competition experiments. Haven's (1973) experiments on gastropods illustrate this type of result. *Acmaea scabra* and *A. digitalis* occur sympatrically in the intertidal off the coast of California. In a series of reciprocal field experiments using fenced enclosures, Haven finds that removal of *A. digitalis* results in increased growth of *A. scabra,* but removal of *A. scabra* usually does not result in increased growth of *A. digitalis.* At this site *A. digitalis* is larger than *A. scabra;* it is unknown whether this same result would be obtained in other populations with different size arrangements of these two species. Haven's results suggest that although interspecific competition does occur between the two gastropods, the effect of the interaction is highly asymmetric.

Other competition studies have suggested a linear competitive hierarchy based on experiments in which only one of the two species was removed. The better competitor in these studies is assumed to occupy the more favorable habitat. These studies provide strong evidence for interspecific competition but the degree to which the interaction is asymmetric cannot be evaluated from these experiments. Connell's (1961a) classic experiments on barnacles in Scotland are an example. Experiments in which *Chthamalus* populations were protected from *Balanus* show that *Chthamalus* survived well in all intertidal areas. In situations in which both species were present, however, *Balanus* prevents establishment of *Chthamalus* in the lower zone. Connell suggests that the upper limit of *Balanus* is probably determined by desiccation (Connell, 1961b), but the reciprocal experiment—removal of *Chthamalus*—was not performed.

Lubchenko's (1980) study of macroscopic red and brown algae in New England rocky intertidal communities is similar in experimental design and outcome to Connell's results. She demonstrates experimentally that the red alga *Chondrus crispus* prevents two *Fucus* species from occupying the low intertidal zone. As a result, the *Fucus* species are restricted to the mid-zone where they grow more slowly than they could in the low zone, in the absence of *Chondrus*. Lubchenko suggests the upper limit of *Chondrus*, in contrast, is set by physical factors rather than by competition.

These studies suggest that asymmetric effects of competition on interacting species are common in natural communities, and the taxonomic range of asymmetric effects of competition is broad. Other experimental studies include starfish (Menge, 1972), hermit crabs (Bertness, 1981), fishes (Hixon, 1980), lizards (Dunham, 1980), insects (Thornhill, 1980), and plants (Grace and Wetzel, 1981). In a compilation of studies on potential insect competition in the field, Lawton and Hassell (1981) conclude that the asymmetric effects of competition between species exceeds symmetric results by a ratio of 2:1. These asymmetries imply that selection pressures resulting from competition are unlikely to be equal on the competing species. The critical question is whether these linear competitive hierarchies are likely to lead to continual coevolution between the competitors, and this question can take the following forms:

1. Should a superior competitor be expected to evolve in response to interactions with an inferior competitor, either by evolving away from the inferior competitor in resource use or by increasing its competitive ability even further?

2. Under what conditions should selection act to increase competitive ability of an inferior competitor rather than act to decrease overlap in resource use with the superior competitor?

An expected possible result of interspecific competition in many mathematical models of two-species interaction is divergence in resource use by both species (MacArthur, 1972; Lawlor and Maynard Smith, 1976; Roughgarden, 1979; Slatkin, 1980). But asymmetries are important in interspecific competition and it is not at all intuitive that both species should diverge in an asymmetric competitive interaction. Instead, it seems more likely that a superior competitor should diverge in response to an inferior competitor only if the cost of winning in competition for a particular resource is greater than the benefit gained from obtaining the resource.

Moreover, the results of some studies suggesting linear competitive hierarchies, including those cited earlier, seem so asymmetric in their effects on the interacting species that the superior competitor may not even be described properly as competing with the inferior competitor; that is, many of these relationships may be effectively amensalistic—affecting negatively one of the species but having little or no negative effect on the other species. Considering only the superior competitor, then, continual coevolution should not be expected often between species exhibiting linear competitive hierarchies.

An inferior competitor can respond evolutionarily to competition by diverging from the superior competitor in resource use or by increasing competitive ability. Divergence can be in the use of habitat or resources in the time that a resource is used, such as in fugitive species that colonize new sites quickly, reproduce, then disperse before superior competitors become established (Hutchinson, 1961; Pickett, 1976). Alternatively, if an inferior competitor responds evolutionarily by increasing its competitive ability—thereby decreasing the asymmetry in the effect of the interaction—then there is a potential for coevolution with the two species continually escalating their competitive abilities over evolutionary time. At least some reciprocal field experiments show reciprocal competitive effects among interacting species, suggesting that some species are close in competitive ability for at least part of their evolutionary history (e.g. McClure and Price, 1975; Inouye, 1978; Smith, 1981; Williams, 1981). The problem is whether this represents just one stage in a seesawing of competitive abilities.

Another way of phrasing this problem is to ask if the relative competitive abilities of species derive from interspecific competition. Laboratory experiments using fly species that differ initially in competitive ability suggest that reversals of competitive ability can occur under these restricted conditions (Pimentel et al., 1965). Gill (1974) argues that increases in competitive ability often result from interspecific competition and take the form of increases in interference mechanisms. He terms this alpha-selection and suggests that exploitative interactions are a transient form of competition that are replaced quickly over evolutionary time by interference mechanisms. Gill, however, does not consider explicitly competitive interactions that are highly asymmetric in their effects.

Although an increase in competitive ability as a result of interspecific competition is an appealing thought, it is not clear that this is a common result of interspecific interactions in natural populations. Moreover, it is not clear that the relative competitive abilities found in interacting species are the result of selection mediated by the interaction. For example, Harger (1972a,b, and earlier), in one of the most detailed studies of competition in the literature, finds that the competitive abilities of two species of sea mussel *Mytilus edulis* and *Mytilus californianus* seem to be mostly fortuitous results of adaptations to cope with the physical environment.

High or increased interspecific competitive ability may also be a fortuitous result of intense intraspecific competition. Abrams (1980) estimates that in the tropical hermit crabs he studied, intraspecific competition is at least an order of magnitude greater than interspecific competition. In the leafhoppers *Eupteryx cyclops* and *E. urticae* on nettle, both intraspecific and interspecific competition decrease female fecundity; but intraspecific competition in experiments with only one species present decreases fecundity more than interspecific competition in experiments with a similar combined density of both species (Stiling, 1980). Relative competitive abilities in both the hermit crabs and the leafhoppers under these conditions are likely to be driven by selection resulting from intraspecific rather than interspecific interactions.

The problem, then, is to separate cause from effect in the evolution of competitive abilities, especially since it is so easy to attribute competitive abilities to interspecific competition. For example, it is simple to conjure scenarios involving interspecific competition in the evolution of arborescence in plants: individuals that grow taller than neighbors receive more sunlight and as a result may be healthier and produce more offspring or perhaps produce them at an earlier age. But this is only one of a number of possible scenarios. Although the arborescent habit may give an individual a competitive advantage in interspecific interactions that could be shown experimentally, this does not imply automatically that the primary basis of selection is interspecific competition. The neighbors may be more likely to be conspecifics rather than other species, and intraspecific competition may be the primary selective force. Alternatively, neither intraspecific nor interspecific competition may be the basis of selection in some instances of arborescence. In a thoughtful discussion of woodiness in plants on oceanic islands, Carlquist (1974) lists a variety of selection pressures that could influence the evolution of arborescence and/or woodiness, including temperature, humidity, and the presence or absence of large grazers.

The general point is that the basis of competitive hierarchies among coexisting species is not necessarily interspecific competition and probably often is not. How both superior and inferior competitors evolve in response to competition demands closer analyses of the asymmetries in the effects of competition on interacting species. Species can respond to interspecific competition in a variety of ways of which divergence of both species is only one possible outcome.

The only experimental field study designed specifically to test whether competing species can coevolve by increasing their competitive abilities is Hairston's (1973, 1980a,b) study of *Plethodon glutinosus* and *P. jordani* in the southeastern United States. These two species overlap in elevation only narrowly in the Great Smoky Mountains but widely in the Balsam Mountains (Hairston, 1951). Hairston (1951, 1973) hypothesizes that the differences in elevational overlap of these species between the two mountain ranges indicate differences in the degree to which the different populations experience competition: narrow elevational overlap indicating high levels of competition and wide overlap indicating low levels of competition.

In a reciprocal removal experiment, Hairston (1980a) removed either one or the other species from a series of plots in both mountain ranges and observed changes in the number of individuals and the number of juveniles added to each study plot over five years. (Five years was predetermined as the time for termination of the experiment, since this corresponds to the minimum estimate for the generation time of *P. glutinosus* and hence the minimum time in which a significant change in the populations could be expected. Data collected on *P. jordani* during the study indicated an eight year generation time for this species.) Removal of the congener resulted in either an increase in population levels *(P. glutinosus)* or an increase in the number of juveniles in the population *(P. jordani)* in both the Smokies and the Balsam Mountains. The responses to removal of the potential competitor, however, were greater in the Smokies than in the Balsam Mountains, as predicted. The competition coefficients, calculated directly from the data, indicated an effect of the interaction on population levels of both species, although the effect was asymmetric. Hairston concludes that interspecific competition occurs between these two species, affects both species, and is greater in the Smokies where the species overlap only narrowly than in the Balsam Mountains.

To test for interspecific competitive ability in these populations, Hairston (1980b) performed simultaneously another set of experiments in which he added individuals of *P. jordani* from each mountain range to study plots in the other mountain range, after removing the resident *P. jordani* individuals. *Plethodon jordani* populations in the two ranges differ in the presence or absence of markings on their cheeks. Therefore, introduced individuals could be separated readily from native individuals. The reciprocal field transplant was not performed because *P. glutinosus* lacks suitable markings to differentiate the populations. The prediction was that *P. jordani* from the Smokies (narrow overlap, high competition) would have a greater effect on *P. glutinosus* from the Balsam Mountains than *P. jordani* from the Balsam Mountains would have on *P. glutinosus* from the Smokies. The results were in accord with the prediction. Hairston concludes that the Balsam Mountains species do not compete greatly and that selection for increased competitive ability has not occurred in these populations. He suggests that these populations may have diverged along some resource gradient. In contrast, he suggests that the populations in the Smokies are coevolving in competitive ability.

These excellent experiments indicate certainly that the populations in the different mountain ranges differ in their interspecific competitive abilities, and it seems clear that these species are competing. The life histories of these species and the relative abundance of other salamanders in these communities also seem to provide a set of ecological conditions that could favor continuing coevolution of these species in some populations. These salamanders are effectively sessile once established, with home ranges of 1.7–11.5 m² for *P. jordani* and 7.5–14.4 m² for *P. glutinosus* (Merchant, 1972). In addition, these two species are by far the most abundant salamander species at the study sites (Hairston, 1980a), and most interactions are likely to be between these two species. The removal experiments did not affect any of the other plethodontid salamanders in the communities (Hairston, 1981). Whether interspecific competition is the only or the major selection pressure generating differences in competitive ability between populations of these two species, however, cannot be determined until more is known about how these populations differ in their use of resources and, especially, in intraspecific interactions that may shape their competitive abilities.

NONHIERARCHIES

Not all competitive interactions are asymmetric, always favoring the same species. Some reciprocal field experiments have demonstrated that in some interactions one species wins on one part of the resource or habitat spectrum and the other species wins on another part of that spectrum (Inouye, 1978; Larson, 1980). Where these interactions involve resources that are mixed as a mosaic such as in flower species visited by bumblebees (Inouye, 1978), or where the habitats are not sharply segregated as in bathymetric segregation in rockfish (Larson, 1980), continual coevolution between the species may result. The probability of encounter between the species may remain high, either through exploitation or interference competition.

Still other competitive interactions are highly variable in outcome depending either upon a mix of environmental variables that are seldom constant or upon which species first colonizes a new area. Some species of coral reef fish seem to interact this way. At least eight species of herbivorous pomacentrid fish live in rubble areas in the Capricorn reefs of the Great Barrier Reef. Sale (1979 and earlier) studied intensively three of these species that utilize similar sites on the upper reef slope. Individuals maintain interspecific territories up to 2 m² in size. Territorial space is a limiting resource for these species as indicated by the rapid rate at which vacated territories are reoccupied (Sale, 1975). The three species that compete for these spaces, however, do not seem to partition the reef slope in any way that would even suggest that the species may have diverged in resource use. Furthermore, these species seem to be near equal in competitive ability (Sale, 1978); when a space is vacated, any of the three species may occupy it (Sale, 1979). Which species gains access to the newly opened territory is unpredictable. The winner in these interactions is the first individual to find the new site, regardless of the species.

Sale (1977, 1978) suggests that coexistence in these species is maintained in a lotterylike manner in which no species can outcompete the others predictably because of certain features of this system. First, vacant space is generated at unpredictable intervals mostly because of predation. Second, the wide dispersal of planktonic larvae away from their parents seems to dissociate the number of available recruits from the number of adults of each species surrounding a vacated space in the reef. Finally, a resident fish always has the competitive advantage over an intruder, no matter what the species. Sale (1977, 1978) notes that this is certainly a nonequilibrium system, and that local extinctions of one of the species should be expected, although any tendency toward global extinction must be slow.

It could be argued that the near-equal competitive abilities of these species may be a continually coevolving trait of these species. The inherent unpredictability in where and when a species wins could keep the interaction together. Analysis of the interactions among these species in a wide range of populations would help clarify the extent to which these three particular species interact wherever they are found. In other parts of coral reefs many fish species interact in a very limited space (Talbot et al., 1978). If these species interact with other species in other sites, then the likelihood that these three species will continually coevolve will depend upon the degree to which the demes of these species are separate rather than routinely connected by dispersal of planktonic larvae. The natural history of these interactions needs detailed study at all levels of scale. Recently, Anderson et al. (1981) argue that, at least in chaetodontid fishes, the process of replacement of sites by new fishes is less stochastic than suggested by Sale for pomacentrids, and that broad geographic patterns of distribution in chaetodontids are discernible.

In general, systems such as the one suggested by Sale in which potentially, (1) only a few species are interacting regularly, and in which (2) the competitive outcomes are unpredictable, or in which (3) resources are thoroughly mixed as a mosaic may be more likely to provide an opportunity for continual coevolution than more predictable linear hierarchies on readily divisible resources. These interactions are likely to be a small subset of the range of competitive interactions that occur between species.

CONCLUSIONS

This chapter has focused on three questions crucial to an eventual theory of the ecological conditions likely to engender continual coevolution between pairs or small groups of competing species: (1) Under what conditions is the probability of encounter between particular species likely to be high and remain high? (2) What kinds of resources are most likely to maintain continued coevolution between competitors? (3) Under what kinds of ecological conditions is selection likely to act to increase competitive ability in species that are inferior in competitive hierarchies?

The major suggestion is that continual coevolution among particular sets of competitors is most likely among species sharing traits or inhabiting communities in which the probability of encounter remains high. These conditions include:

1. Species whose individuals have a poor ability to choose sites in which to live and are restricted in movement once established at a site.
2. Species in communities low in numbers of potentially competing species (Connell, 1980).
3. Species that compete for a mutualist that must visit several to many host individuals during its lifetime.
4. Competitive interactions in which the outcome remains unpredictable because environmental conditions change unpredictably, new colonization sites become available unpredictably, and/or critical resources are distributed in a mosaic that is not readily divisible.

These suggestions are meant as hypotheses rather than as conclusions. It is clear that some species compete. If a general theory of the interaction structure of communities is ever to develop in evolutionary ecology, however, patterns in where and when competition leads to continual coevolution among sets of species must be teased out of the many kinds of competitive interactions that occur between species. The need for long-term reciprocal field experiments on competition between species in a wide variety of natural communities is among the most important needs in evolutionary ecology. Moreover, studies comparing the same pairs of species across a gradient of communities varying in richness of competing species, will be the most valuable.

CHAPTER

4

ANTAGONISM AND MUTUALISM

Mutualisms between species encompass an odd conglomeration of interactions in a world often viewed as "red in tooth and claw." Although a wide variety of mutualisms has been described, we still have only a rudimentary understanding of the selection pressures, life history traits, and community attributes that favor the origination and maintenance of these interactions. This lack of a solid understanding of the ecological conditions favoring mutualisms is reflected in the literature on mathematical models of mutualisms. Some mathematical and conceptual models conclude that mutualisms should be expected to be less common in nature than antagonistic interactions, because the models of mutualism often exhibit high degrees of instability (e.g. May, 1973; Van Valen, 1973; Goh, 1979). Other mathematical models of mutualism in coevolving species sometimes exhibit stability depending upon the parameters in the models (e.g. Roughgarden, 1975; Levin and Udovic, 1977; Vandermeer and Boucher, 1978; Heithaus et al., 1980; Addicott, 1981). The most reasonable interpretation of the attempts at modeling mutualisms over the past decade is not that mutualisms are uncommon—the empirical evidence is to the contrary—but rather that we know little about the ecological bases for mutualisms relative to antagonistic interactions. I think both modelers and field ecologists would agree with this interpretation. In the book *Theoretical Ecology,* May (1976, 1981) writes that he would have liked to include a chapter on mutualism much like the chapters in the book on arthropod predator–prey systems, herbivore–plant systems, and competition and niche theory. He did not because he thought that both the theoretical and empirical bases of mutualism were still insufficiently known to be summarized as a separate chapter.

Nevertheless, there are several fronts on which the theory of mutualism is advancing and some initial syntheses are emerging. This chapter and Chapters 5 and 6 suggest several patterns and hypotheses on the selection pressures that favor the origin and maintenance of mutualisms between species. The arguments focus on the following ecological questions:

1. How are antagonistic and mutualistic interactions related evolutionarily? (Chapter 4)
2. What is the basic unit of interaction in mutualisms? (Chapter 4)
3. Under what life history and habitat conditions is the probability of a truly mutualistic encounter between potential mutualists likely to be high, thereby allowing ongoing coevolutionary change between particular sets of mutualists? (Chapter 5)
4. How do mutualisms differ in the extent to which species become obligately committed to the interactions over evolutionary time? (Chapter 6)

All of these questions are concerned with the search for patterns in how different kinds of mutualism contribute to the interaction structure of communities and how these interactions are held together over evolutionary time. By design these questions are similar to those addressed in the chapters on antagonistic trophic interactions and competition.

In this chapter I consider the evolutionary relationships between antagonistic and mutualistic interactions. The general theme of the chapter is the following assertion: the richness of mutualisms in communities depends upon the richness of antagonistic interactions. By this assertion I do not mean that all mutualisms depend upon antagonistic interactions to be favored by selection; Chapter 5 considers other selective bases for the evolution of mutualisms. But such a wide array of mutualisms have their evolutionary origins in antagonistic interactions that it seems worthwhile to devote a chapter to the evolutionary relationship between antagonism and mutualism as a background for the conditions favoring coevolution of mutualists developed in the following chapters.

The evolution of mutualisms depends upon antagonistic interactions in the following ways: (1) many mutualisms derive evolutionarily from initially antagonistic interactions between species, and (2) many other mutualisms involve a unit of interaction encompassing at least three species or groups of species, including an antagonistic pair of species and a mutualistic pair of species. Under both conditions mutualism represents an evolutionary result that is dependent on selection driven by an antagonistic interaction.

MUTUALISM FROM ANTAGONISM

Classic heroic epics and fairy tales are easy to follow partly because the intentions of the characters are often purely good or purely evil. The inevitable additional character, caught between the two extremes, is more complex but his or her actions are understandable because the extremes are well defined. If this were also true of the interactions in biological communities, we could then list all of the species in a community, draw lines between them indicating interactions, then label those lines with pluses, minuses, or zeroes, depending upon the type of interaction each represented. Hence all lines from seed-predators to plants would get minuses and all lines from fruit-eaters to plants would get pluses.

Instead, the interactions between species are built on entirely different rules. All the interactions are inherently selfish. Whether any of these selfish interactions is beneficial, detrimental, or inconsequential for the fitness of the affected individual depends upon the relative gain to loss in fitness that the interaction produces. Therefore, we should expect to find continua of antagonism to mutualism in superficially similar interactions among species and even in the same interaction among different populations of the same species, pairs, or groups.

The theoretical problem in understanding the evolution of mutualistic interactions from antagonistic interactions is deciphering the ecological conditions that favor selection toward mutualism rather than toward escalation of defenses and/or avoidance of the interaction. I suggest that a change in outcome from antagonism to mutualism is most likely in interactions that are inevitable within the lifetimes of individuals.

The reasoning follows from the considerations of defense reactions to inevitable interactions developed in Chapter 2. If it is unlikely that individuals can avoid a specific antagonistic interaction, then selection will favor individuals that have traits causing the interaction to have at least less of a negative effect on them. This selection regime sets the stage for the evolution of the interaction toward commensalism or mutualism. Although most of these interactions probably do not lead to coevolved mutualisms, some do generate mutualisms. Hence mutualisms can derive from a change in outcome in inevitable antagonistic interactions between particular species and may have their evolutionary origin in defense reactions of species.

Pollinators and Plants

The evolution of interactions between plants and pollinators provides some of the clearest examples of change in outcome of interactions from antagonistic to mutualistic. The habit of feeding on spores is among the oldest in terrestrial arthropods (Kevan et al., 1975), and there is general acceptance among biologists that the early insect pollinators of angiosperms fed on pollen, ovules, and/or seeds and other flower parts (e.g. Leppik, 1975; Crepet, 1979). Undoubtedly, the vast majority of these interactions were detrimental to the plants, and the closed carpels of angiosperms were probably a defense against these flower visitors (Mulcahy, 1979). Nonetheless, these antagonistic interactions provided a basis on which selection could act. Some of these flower visitors were less detrimental to flower parts than others and some plants possessed floral traits that caused the interaction to be less detrimental to the plant, and at some point in time, actually beneficial.

The large-scale transition point between antagonism and mutualism, favoring pollination by insects over pollination by wind over large geographic areas, may have been a consequence of changing climates selecting against wind-pollination rather than strictly selection for insect–plant mutualisms as an absolutely superior mode of pollination. The rise to worldwide dominance of angiosperms is correlated with the mid-Cretaceous expansion of epeiric seas (Axelrod, 1970). The expansion of these seas provided climatic conditions that were warm and moist over large geographic areas. Raven (1977) argues that under these conditions primitive angiosperms that were pollinated by insects would have been favored by natural selection over wind-pollinated plants since pollen does not travel well by wind in moist tropical environments. Therefore, because of changing climatic conditions, the benefits of insect visitation increased relative to the loss of pollen and ovules to the visitors.

The development of floral nectaries further reduced the cost of the interaction relative to the gain. Although some pollinators such as bees (Baker and Baker, 1979) still rely heavily on pollen as the reward for visits, the evolution of floral nectaries transferred some of the reward away from the plant reproductive structures themselves and likely lessened the cost of the interaction.

The interactions between agaonid wasps and figs and between incurvariid moths and yuccas illustrate more specifically the evolutionary transition from antagonism to mutualism. Pollination in figs is accomplished by agaonid wasps that oviposit in some of the ovules of the inflorescences that they pollinate. In four fig species studied by Janzen (1979) in Costa Rica, 41–77% of seeds were killed on average by the aganoids and associated parastic Hymenoptera. These percentages can be considered as direct costs of the mutualism to the plant. The interaction is simultaneously antagonistic and mutualistic. There seems to be general agreement among those studying agaonid–fig interactions that the phylogenetic affinities of the agaonid wasps indicate evolution of the interaction from a pre-agaonid wasp that was parasitic in the flowers, although there is disagreement on the precise floral parts fed upon by the pre-agaonids (Galil, 1973; Ramírez, 1976; Wiebes, 1979).

The interactions between incurvariid moths in the genus *Tegeticula* and yuccas are similar to agaonid–fig interactions. Two species have been studied in detail: *T. yuccasella* (Riley, 1892; Trelease, 1893; Davis, 1967) and *T. maculata* (Powell and Mackie, 1966; Aker and Udovic, 1981). As in agaonid–fig interactions, a female *Tegeticula* collects pollen that she deposits on a stigma after she oviposits into the pistil (although in agaonid–fig interactions this is not always an active process on the part of the wasp). By this process the larvae are guaranteed developing seeds on which to feed, and the plant is provided with an effective pollinator, but at a cost of some seeds. The cost of this interaction to the plant in lost seeds is not known.

The genus *Tegeticula* represents one genus among several that comprise the subfamily Prodoxinae, which are borers in various parts of agavoid plants, and the evolutionary transition from antagonism to mutualism is indicated within the subfamily. Davis (1967) suggests that the progenitor to the subfamily was a group with fruit-boring larvae (Figure 4.1). Most of the species in the subfamily feed as larvae by boring in the fleshy parts of the agavoid fruits or in flower stalks. The three species of *Tegeticula* are simply borers that develop as larvae on seeds rather than in the fleshy portion of fruits or the flower stalks. Ancestral *Tegeticula* that happened to pollinate the ovules as they oviposited undoubtedly produced more young than those that did not. *Tegeticula maculata* continues to oviposit occasionally into developing fruits rather than into unfertilized ovules (Aker and Udovic, 1981), so the relationship is not always mutualistic. The single species in *Parategeticula* may also pollinate its host plant, but the biology of this species is known incompletely (Davis, 1967). The evolutionary relationship between antagonism and mutualism is particularly clear in these interactions because the feeding habits of the related genera provide an indication of how the evolutionary transition in the interaction may have occurred.

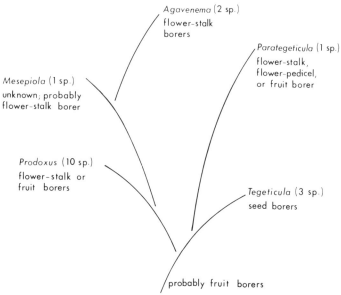

Agavenema (2 sp.)
flower-stalk
borers

Parategeticula (1 sp.)
flower-stalk,
flower-pedicel,
or fruit borer

Mesepiola (1 sp.)
unknown; probably
flower-stalk borer

Prodoxus (10 sp.)
flower-stalk or
fruit borers

Tegeticula (3 sp.)
seed borers

probably fruit borers

FIGURE 4.1.   Suggested phylogeny for the subfamily Prodoxinae (Lepidoptera: Incurvariidae) and the feeding habits of larvae in each genus. Adapted from Davis (1967).

Vertebrates and Seeds

There are parallels in the transition from antagonism to mutualism in interactions between plants and flower visitors and interactions between seeds and vertebrates. In interactions involving pollen, nutlets, or cones as the reward, the dispersal unit itself (pollen, seeds) is eaten by the mutualist. The mutualism is based on the subset of pollen grains or seeds not eaten. In contrast, in interactions involving nectar or fleshy fruits as the reward, a separate new resource is offered to the mutualists. The evolutionary ties of these interactions to previous antagonistic interactions become less apparent than where the dispersal unit itself is eaten.

The relationship between antagonism and mutualism is evident in the combined studies on seed predation and seed dispersal, which merge in the studies of predator (or parasite) satiation and seed caching by vertebrates. Seed dispersal via caching by animals depends upon the evolutionary maintenance of predator satiation (mast seeding) as a defense mechanism to increase the probability of survival in seeds. If few seeds are produced locally or regionally within a year, seed predators and parasites are likely to eat most of the seed crop; whereas if many seeds are produced, some seeds may escape through satiation of the seed predators and parasites (Janzen,

1971a, 1974a; Forcella, 1980; Silvertown, 1980). Furthermore, if the seed predators live in environments where (1) food shortages occur at some season of the year and (2) seeds can be safely cached without rotting until the season of food shortage, then the seed predators may cache some of the seeds for later use (Janzen, 1971a, 1974a; Snow, 1971; Roberts, 1979). Some of the cached seeds may not be recovered by the seed predators, and instead germinate (Vander Wall and Balda, 1977; Ligon, 1978; Stapanian and Smith, 1978; Bossema, 1979). Therefore, the conditions favoring the evolution of seed caching mutualisms are a subset of the conditions favoring mast seeding as a defense mechanism, and in this way antagonism and mutualism are linked closely in these interactions.

The critical constraint on mast seeding as a form of defense is that the numerical and functional responses (sensu Holling, 1961) of the predators or parasites must be lower than if reproduction were distributed more predictably in time or space. The only critical characteristics for predator satiation to be effective are large numbers of offspring produced synchronously, unpredictability in when or where reproduction takes place, and short duration for the reproductive period. Previously, Janzen (1971a) argued that partial to total host- or prey- specificity in the predator or parasite was also essential, but he later relaxed this requirement when he considered the evolution of mast seeding in bamboo species (Janzen, 1976a). The requirement of host or prey specificity also does not hold for periodical cicadas, whose adults are attacked by both a specialized fungus and by a variety of less prey-specific insectivorous vertebrates, and for whom the term predator satiation was coined (Lloyd and Dybas, 1966).

The conditions for predator satiation to be effective will most likely be met in habitats in which the diversity of ecologically related host or prey species is relatively low (Janzen, 1974a). Under these conditions the community is comprised of few species with large populations that can satiate the predators or parasites. Therefore, predator satiation with population synchrony in reproduction is expected to be ineffective as a defense mechanism in species-rich communities with populations of widely spaced individuals (Janzen, 1974a). In fact, plants that reproduce synchronously and periodically within populations occur mostly in temperate zones in habitats in which population levels of individual tree species are high (Svardson, 1957; Smith, 1970; Ligon, 1978), in species-poor dipterocarp forests of southeast Asia (Janzen, 1974a), and in many bamboo species that occur as large even-aged populations (Janzen, 1976a). Moreover, Janzen (1974a, 1976a) argues that periodic synchronous fruiting is more likely to evolve in temperate forests and in dipterocarp forests than in species-rich tropical forests because populations of seed predators are depressed each winter in temperate forests and because dipterocarp forests occur on poor soils capable of supporting only very low populations of animals. In these environments the probability is highest that the plants could satiate predators and parasites in a mast year.

In general, then, the evolution of predator satiation as a means of defense during reproduction is likely to be limited to perennial species that live in habitats with low diversity of ecologically related species and are unpredictable in either time or place of reproduction. Host-specific predators or parasites must evolve, in response, either dormancy patterns keyed into the same environmental cues as the plant, a less host-specific feeding habit to survive food-poor years, or high mobility. Specialized seed-eating birds roam widely in years when seed crops in their local areas are low, both in Europe (Svardson, 1957) and in the southwestern United States (Vander Wall and Balda, 1977; Ligon, 1978).

For the arguments developed here, it does not matter if the initial selection pressure for mast seeding is the potential mutualist vertebrate seed cachers or other seed predators or parasites. In either circumstance—and it is unlikely to be an either/or phenomenon—the mutualism relies on the maintenance of the initial defense reaction of mast seeding. This defense reaction sets up the ecological conditions under which more seeds are produced than are needed immediately by the vertebrates and can be cached for later use.

Seed-caching mutualisms are, therefore, limited geographically to habitats where the ecological conditions permit mast seeding as a successful means of defense against seed predators or parasites and the mutualist vertebrates vary with the traits of the seeds required for seed caching to be successful as a means of dispersal. Seed caching by birds occurs primarily in temperate plant species (Roberts, 1979), especially between some species of corvids and some oaks and pines. Clark's Nutcracker relies heavily on and is an effective dispersal agent of *Pinus abicaulis* (Tomback, 1978) and *Pinus edulis* (Vander Wall and Balda, 1977). Pinyon jays also rely heavily on *Pinus edulis* and are also important dispersal agents of this species (Ligon, 1978). Both species of pine share characteristics that Smith and Balda (1979; see also Balda, 1980; Vander Wall and Balda, 1981) associate with selection for avian dispersal in pines: relatively large seed size, absence of wings on seeds, lack of quick release of seeds from cones, indications of seed quality on the seed coat, positioning of cones on trees to maximize visibility, lack of spines on the cone scales, and relatively thin seed coats. Smith and Balda call these "enticer species" to differentiate them from nonenticer wind-dispersed species that have the opposite characteristics.

Dispersal of oaks involves a broader range of vertebrate taxa. Both jays and squirrels store acorns and other large hard seeds and these seeds seem adapted to this means of dispersal (Smith, 1975; Stapanian and Smith, 1978; Bossema, 1979). European jays *(Garrulus g. glandarius)* depend heavily upon acorns and individuals may plant several thousand acorns individually in the ground in October; many of these germinate later (Bossema, 1979). Unfortunately, the relationships between the now extinct passenger pigeon and oaks and beechnuts in North America can never be unravelled (Schorger, 1955). In the tropics, caching appears to be primarily by rodents (Smythe, 1970). Roberts (1979) suggests that birds may be excluded from such behavior in these regions because successful caching must involve relatively hard large seeds to prevent rapid decay, putting such seeds outside the limits for most avian bills.

The theoretical importance of seed-caching mutualisms is that, like the interactions between figs and agaonid wasps and between yucca and *Tegeticula,* the relationship between antagonism and mutualism is readily apparent. By comparison, evolution of fleshy fruits provided an alternative route to vertebrate – seed mutualisms where the antagonistic component of the interaction has become less apparent. The term fleshy fruit is used here in its ecological sense: a soft covering around seeds that attracts dispersal agents who eat the covering and discard the seeds (Thompson and Willson, 1979). As with the evolution of nectaries in flowers, fleshy fruits represent a new resource offered to the mutualist; this new resource must certainly lower the cost of the interaction for some plants, since potentially all seeds could survive the interaction in this type of mutualism. Also, fruits could be less costly to the plant as a reward for mutualists, since the fruits need not contain all the nutrients necessary for seed survival. The interaction evolves as a change in outcome from earlier vertebrate – seed interactions.

The evolution of fleshy fruits, however, does not imply a multistep evolutionary pathway in which cached seeds is an intermediate step and production of fleshy fruits is the final step. Rather these types of seed dispersal may represent responses to different selection pressures in the development of mutualisms from antagonistic interactions. Some of the differences in these types of vertebrate – seed mutualisms are suggested in Table 4.1. The evolution of mutualisms involving birds and fleshy fruits is considered in detail in Chapter 6.

Cleaner Fishes and Host Fishes

The relationship between antagonism and mutualism is also evident in the interactions between cleaner fish and host fish. Marine and freshwater fish in a number of families are known to feed on ectoparasites and other materials on the body surface and oral and gill cavities of host fishes (Gorlick et al., 1978). The cleaner fish gains food and the host fish is

Table 4.1.
General Differences in Vertebrate−Seed Mutualisms[a]

	Reward to Vertebrate	
	Seeds	Fleshy Fruits
Dependence on predator satiation	High	None[b]
Seeds planted by animal	Yes	No
Ability of reward unit (seed, fruit) to survive caching	High	Low
Selection exerted by animal on seed size	Large seeds	Large or small seeds
Plant growth form	Trees, usually	Trees, shrubs, vines, herbs

[a]Based on whether the reward unit is the seed itself or a fleshy fruit.
[b]Some interactions between seeds and fleshy fruits seem to vary with the size of the fruit crop, but the basis of selection does not seem to be related to predator satiation (e.g. Howe and Vande Kerckhove, 1979).

generally considered to gain increased probability of survival through the interaction (Limbaugh, 1961). Obligate cleaner fish are alleged to share several traits: (1) obligate dependence of the cleaners on ectoparasites and tissues of host fish for nutrition, (2) restriction to specific sites for long periods of time, (3) occurrence as single individuals or pairs, (4) defense of sites against conspecifics, and (5) overall freedom from predation (Itzkowitz, 1979).

There are, however, still few data on the actual degree of attraction of host fish to cleaner fish and the effect of the interaction on host fish fitness. The blanket suggestion of mutualism for these interactions has been criticized by several authors (Hobson, 1969; Losey, 1971; Gorlick et al., 1978), and the experimental work has produced some conflicting results on effects of cleaner fish on ectoparasite population levels (Gorlick et al., 1978). These conflicting results may indicate that these interactions vary in position along the continuum from antagonistic to mutualistic interaction. Gorlick et al. (1978) suggest that at least some cleaner fish evolved from species that fed on scales and mucus of host fishes. Such fishes exist today and do not generally ingest ectoparasites (Major, 1973). Hence the interactions and their effects on the host will tend to vary with the species of cleaner and host and the ecological context of the interaction (Gorlick et al., 1978; Losey, 1979).

In summary, the interactions between pollinators and plants, vertebrates and seeds, and cleaner fish and hosts serve to illustrate the close evolutionary dependence of many kinds of mutualism on pre-existing antagonistic interactions. In all of these kinds of interaction, the evolution of traits changing the outcome of the interaction from antagonistic to mutualistic is likely only when the interactions have a high probability of occurring within the lifetime of each individual. Attractants for pollinators and seed dispersers are costly to plants and are unlikely to be produced when the interaction is much less than inevitable; host behaviors allowing carnivorous fish to approach a host unharmed could evolve only if the interactions between potentially mutualistic cleaner fish and hosts have a high probability of occurring and can be maintained by selection. Therefore, antagonistic interactions that are avoidable by hosts or prey are unlikely to evolve toward mutualism through a change in outcome from antagonistic interactions, whereas inevitable antagonistic interactions provide the evolutionary raw material for coevolved mutualisms.

This does not mean that all inevitable antagonistic interactions lead to coevolved mutualisms, but only that these interactions provide the potential ecological conditions for the evolution of mutualisms. Some recent papers have suggested potentially beneficial effects of herbivores on plant growth (Stenseth, 1978; McNaughton, 1979; Dyer, 1975; Owen, 1980; Owen and Wiegert, 1981) but whether any of these actually involves a mutualism where the fitness of both herbivore and plant is increased through the interaction remains to be demonstrated. Some other proposed effects of herbivory on plants may generate odd beneficial effects on the plant, such as the potentially increased ability of red mangrove (*Rhizophora mangle*) to survive wave action because attack by wood-boring marine isopods increases branching of aerial prop roots (Simberloff et al., 1978). These and related suggestions (Owen and Wiegert, 1976, 1981; Petelle, 1980) all indicate ways in which mutualism may derive potentially from antagonism and are important in generating flexibility in thinking about the possible evolutionary outcomes of interactions. Only careful cost and benefit analysis of the effects of the interactions on individual plants, however, can suggest whether any of these interactions represent actual mutualisms.

THE UNIT OF INTERACTION

Most studies of mutualism consider directly the interaction between two species, yet the evolutionary unit of many mutualisms involves at least three species in a way that emphasizes the evolutionary relationships between antagonism and mutualism. This point is obvious in the context of protection of one of the species from parasites, grazers, predators, or competitors by another species. *Pseudomyrmex* ants protect acacias from both herbivores and encroaching vines (Janzen, 1966); the shrimp and crab symbionts of the coral *Pocillopora elegans* protect their host from the predatory sea star *Acanthaster planci* by snipping at the sea star's spines and tube feet (Glynn, 1980). These, however, are not the only contexts in which the unit of interaction involves at least three or more species or groups of species. This section considers a variety of mutualisms involving more than two species in the hope of providing a stimulus for studies designed to analyze the whole unit of interaction in mutualisms rather than a part of the unit, and also to emphasize another way in which mutualisms are often dependent evolutionarily on antagonistic interactions. The general point is that, although some mutualisms involve mostly a change in an interaction between two species from antagonism to mutualism, other mutualisms are built on interactions involving at least an antagonistic pair of species and a mutualistic pair of species.

Some adult insects carry other mutualistic species to their breeding sites at which the phoretic individuals unwittingly provide protection for the offspring of the insect by lowering defenses of the insect's host or by killing competitors of the insect. Springett's (1968) study of burying beetles *Necrophorus* spp. and the mite *Poecilochirus necrophori* on the Farne Islands is a clear example. *Necrophorus* beetles search out and bury mouse corpses on which they rear their young. Several beetles are often involved in burying a corpse, but it is eventually monopolized by one pair of beetles. The larvae develop on the corpse and the female beetle feeds the young at the beginning of each instar. *Necrophorus* adults commonly carry 10–23 mites per individual. The mites feed on *Calliphora* fly eggs and larvae that are on the mouse corpses; these flies are the major competitors of *Necrophorus* for the corpses (Elton, 1966).

Springett analyzed the interactions by setting up a series of laboratory experiments with wood mice, *Apodemus sylvaticus,* corpses, and various combinations of flies, beetles, and mites. Only in cultures lacking fly eggs did the beetle larvae survive. This occurred when either no flies were initially added to the culture or when flies were added but so were mites. A similar set of experiments in which 3-mm fly larvae were introduced rather than fly eggs resulted in no successful survival of the beetles. Neither the beetles nor the mites are capable of killing fly larvae over 5 mm, indicating the importance of the mite's presence soon after the discovery and burying of the corpse by the beetles.

The interaction seems to be truly mutualistic. The mites are dispersed to new sites with available food by riding phoretically on the adult beetles and the beetle gains protection from the flies. Springett suggests that the relationship is probably obligatory. D. S. Wilson (pers. comm.) has obtained similar results in field trials in northern Michigan and has identified morphological traits on the bodies of the beetles that seem to function specifically as sites for attachment by the mites and behaviors that allow for mite attachment.

Parasites, grazers, and predators often form mutualisms with symbionts that allow host or prey species to be attacked or to be utilized more effectively by increasing the availability of nutrients or by detoxifying defensive compounds in the host. The basic unit of interaction is not a two-species interaction between the herbivore and plant or the parasite and its animal host, but rather a three-species interaction in which a symbiont species or group of species is a critical link. It is true certainly that a herbivore and its gut symbionts are often so inextricably linked that they function effectively as an evolutionary unit, but assessment of the evolutionary basis of either the mutualistic or the antagonistic component to these interactions must involve ultimately all three species or species groups.

Examples of these kinds of interaction are widespread. Ruminant mammals harbor a highly complex group of gut symbionts (McBee, 1971; Hungate, 1975). Aphids, which feed on phloem poor in some nutrients, harbor symbionts that synthesize these compounds (Dadd, 1973; Houk and Griffiths, 1980). The ichneumonid wasp *Campoletis sonorensis* harbors a virus that reproduces in the ovary of the wasp. When the wasp oviposits into the larvae of the tobacco budworm *Heliothis virescens* the virus is also transported into the host. The presence of the virus suppresses the host's defensive response to encapsulate the egg of the parasitoid. In the absence of the virus the parasitoid is encapsulated and dies (Edson et al., 1981). The evolutionary unit of this interaction involves the parasitoid, the virus, and the budworm.

Still other mutualistic interactions involve a basic evolutionary unit of more than three species or groups of species. In many of the tropical and subtropical plant taxa occupied and considered to be defended by ants, the unit of interaction involves at least four species or groups of species: the plant, the ants, the herbivores or vines that the ants keep away, and also Homoptera that are tended on the plants by the ants. Naturalists have long known that many ant-inhabited plants in the tropics also harbor Homoptera, including the African *Acacia* species (Ule, 1906; Hocking, 1970). Bailey (1922) notes that all African myrmecophytes with which he is familiar regularly contain scale insects except for one species. As early as 1874, Belt notes that the *Azteca* ants on *Cecropia* species in Central America tend scale insects. Nevertheless, the homopteran component has never been analyzed fully for any obligate ant–plant interaction of which they are a part. Hocking (1970, 1975) suggests that the swollen stipular thorns of African acacias originated as homopteran induced galls through genetic assimilation (sensu Waddington, 1975). Although many evolutionary biologists would argue with Hocking's use of genetic assimilation to explain the evolution of the African ant—acacia interaction, the overall point is that the homopterans may be critical in any ecological assessment of how the interactions between the ants and the plants have evolved.

Perhaps the most graphic example of how mutualisms are often part of a unit of interaction that involves three or more species is Smith's (1968, 1979) study of brood parasitism by the giant cowbird *(Scaphidura oryzivora)* on two host species, the chestnut-headed oropendula *(Zarhynchus wagleri)* and the yellow-rumped cacique *(Cacicus cela)*, in Panama. The interactions between the hosts and the brood parasite are interpretable only in the context of the distribution of colonies of stinging and biting bees and wasps and the probability of attack on oropendula and cacique chicks by botflies in the genus *Philornis*.

The giant cowbird lays its eggs almost exclusively in the nests of colonial nesting oropendulas and caciques. Five types of eggs are laid by *Scaphidura* females: three oropendula mimics, a cacique mimic, and a nonmimetic generalized icterid type. Females laying mimetic eggs generally place only one egg in a nest and are cryptic in their egg-laying behavior; females laying nonmimetic eggs usually lay two to three and sometimes as many as five eggs per nest and are aggressive toward the host females. Whether the eggs are mimetic or nonmimetic varies between host colonies and depends upon the behavior of the host birds. In some colonies host females toss nonmimetic eggs out of the nest, so that only mimetic cowbird eggs survive. In experiments in which egg models of various types were placed in nests, colonies comprised of "discriminators" would accept eggs only if they fell within the narrow range of variation peculiar to that colony. In other colonies, however, females would accept eggs of a wide variety of size, color, and patterns. Smith calls these hosts "nondiscriminators."

Whether a colony is comprised of discriminator or nondiscriminator birds depends upon the presence or absence of hymenopteran nests near the colony. Discriminator colonies are clustered around nests of stinging wasps or nonstinging, but biting, bees. Oropendula and cacique chicks in nests away from hymenopteran nests are parasitized heavily by botflies. The female botflies lay eggs or young larvae on the chicks and chicks with more than ten of these ectoparasites usually die. Why oropendula and cacique nests near these hymenopteran nests suffer lower levels of botfly parasitism is unknown (Smith, 1980). The hymenopteran nests also probably provide some protection against vertebrate predators. Birds under the protective sphere of hymenopteran nests reject cowbird eggs unless they are very much like their own eggs.

Oropendulas and caciques away from hymenopteran nests, however, accept cowbird eggs, and cowbird eggs in these colonies are nonmimetic. Smith finds that under these conditions cowbird chicks in the nest actually increase the probability of host chick survival to the fledgling stage over nests without cowbird chicks. The cowbird chicks eat the botfly larvae from the bodies of the host chicks, so this major source of mortality among chicks is eliminated. The probability of rearing at least one chick to fledgling stage is about three times greater if the other chick in the nest is a cowbird rather than a sibling.

In general, parasitism by cowbirds is disadvantageous to hosts in colonies associated with hymenopteran nests and advantageous to hosts in colonies away from bees and wasps. An eight-year experiment indicated that the response by host birds is learned rather than inherited as an innate response. During the experiment, Smith (1979) switched female chicks 8–17 days old from colonies exhibiting one type of behavior to colonies exhibiting the opposite behavior or, as controls, the same behavior. Only females were used since only females are involved in discriminating behavior and nest activities as adults. In all, 1476 chicks were switched between nests. The relatively few females that bred as yearlings all displayed nondiscriminating behavior, regardless of the behavior of their real or foster mothers. Most two-year old and older birds returned to their foster colony and assumed the behavior appropriate to the colony. In this interaction, then, the mutualism is variable between populations of the same species and depends clearly upon several kinds of antagonistic interaction: (1) the interaction between the botflies and the caciques and oropendulas, (2) the interaction between the hymenopterans, botflies, and perhaps vertebrates, and (3) the interaction between the cowbirds, caciques, and oropendulas that varies between antagonism and mutualism.

Even some of the mutualistic interactions considered in the previous section as deriving from inevitable antagonistic interactions through a change in outcome in the interaction may involve a unit of interaction larger than two groups of species. The degree to which some interactions between birds and fruits or insects and flowers are truly mutualistic can depend upon interactions other than those between the mutualists. If a frugivore remains in a single fruiting tree for a major part of a day eating fruits and regurgitating or defecating the seeds beneath the tree, the basis of the interaction from the plant viewpoint is thwarted. If fruiting trees also attract predators of frugivores, however, then frugivores may be more likely to eat a few fruits at a time and process them away from the fruiting tree. Howe (1979) argues that this positive effect of predators of frugivores on seed dispersal is most likely to apply to bird–fruit interactions involving small frugivores that forage singly or in noncooperative assemblages, in habitats that offer suitable refuges from predators from fruiting trees.

The extent of outcrossing in some trees with large floral displays may depend on the presence of several flower-visiting species that vary in their abilities to monopolize flowers on parts of trees. Frankie, Opler, and Bawa (1976) studied the pollination of a self-incompatible neotropical papilionaceous tree, *Andira inermis*, that is pollinated by solitary bees. They collected 70 species of bees from *Andira* flowers. By using a mark-and-recapture technique, they find that most bees remain on a single tree but between 0.3−3.8% of bee movements are between trees. It is this small percentage that effect cross-pollination. Of the 70 species of bees, however, only six species moved between trees. Why any of the bees should move from tree to tree is a matter of speculation. Frankie et al. suggest that the territorial behavior of some species may drive others away. Also the high activity levels at the trees may drive some of the less aggressive bees to abandon one tree and look for another tree of the same species.

Although the clearly mutualistic bee species in this interaction are the six species that tend to move between trees, the extent to which they move between trees rather than remain on single trees may depend importantly on the presence of the aggressive bees that stay at single trees. The question is if the aggressive "thieves" were removed from the system, would the level of outcrossing remain the same or actually decrease?

The general point of these examples is that the unit of interaction in many mutualisms often involves at least three species, combining a mutualistic component with an antagonistic component. Many mutualisms are evolved in specific response to antagonistic interactions between a pair of mutualists and a third species. Other mutualistic interactions such as the pollinator−plant and bird−fruit interactions discussed above are unlikely to be evolved in specific response to a third, antagonistic species, but the degree of mutualism can depend upon the presence or absence of an antagonist. Heithaus et al. (1980) show that predators or competitors of one of the mutualists can stablize mutualisms in mathematical models.

The implication of these arguments and those in the previous section on the evolution of mutualism from antagonism is that the richness of most kinds of mutualistic interactions in communities depends in a very real way upon the richness of antagonistic interactions. This was the assertion on which this chapter began. It is unlikely that communities poor in their variety of antagonistic interactions will support a rich variety of mutualisms. The major exception is nutrient-gathering mutualisms involving sessile organisms in nutrient-poor environments, e.g. ant-fed plants on white-sand soils, mycorrhizae, and perhaps lichens. These mutualisms are considered in Chapter 5 together with an analysis of life history traits and habitat conditions that make the evolution of mutualisms likely.

CONCLUSIONS

The major arguments of this chapter are that (1) the richness of mutualistic interactions in communities depends evolutionarily upon the richness of antagonistic interactions because (2) many mutualisms derive from inevitable antagonistic interactions through a change in outcome of an interaction, and (3) many other mutualisms are built on a unit of interaction with at least three species, involving an antagonistic pair and a mutualistic pair. This tight linkage between antagonism and mutualism suggests an overall constraint on the development of the interaction structure of communities over evolutionary time. This constraint suggests why Skutch's (1980) wish that evolution had proceeded more along a path of cooperation than antagonism could not have been realizable. Antagonism and mutualism are not separate and alternative evolutionary directions.

Mathematical or conceptual models of mutualisms that fail to include the antagonistic components of mutualistic interactions are unlikely to mirror most actual mutualisms. Similarly, field studies of mutualisms need to include both the benefits and the costs of mutualisms in order to decipher how the net benefit is achieved in an interaction. This is important especially if we are to understand how selection shapes mutualisms over evolutionary time. The selection pressures shaping mutualistic interactions act not only to increase the probability that individuals encounter mutualists but act also to decrease the cost of the interaction to individuals.

CHAPTER

5

LIFE HISTORIES, HABITAT, AND MUTUALISM

Chapter 4 suggested several ways in which mutualisms are tied evolutionarily to antagonistic interactions. Having developed this theme, it is important to make the following points at this juncture:

1. Not all antagonistic interactions lead to mutualisms. That is, although many mutualisms derive from antagonistic interactions, this statement differs importantly from any suggestion that all antagonistic interactions lead to mutualisms.

2. Some mutualisms are based on physical stresses rather than on antagonism with other organisms. Whether a mutualism develops between species through antagonism or physical stress (or a combination of these conditions) depends upon life history traits of the interacting organisms and features of the habitat in which they interact.

Therefore, this chapter considers general aspects of life histories and habitats that favor the evolution of mutualisms, plus one specific aspect of life histories—the richness of social behavior—that influences the probability of mutualism.

INTERMEDIATE SURVIVAL AND DISTURBANCE

Roughgarden (1975) made a pioneering attempt to isolate components of life histories that predispose organisms to mutualistic interactions.

76

Through a simple cost-benefit model, he argues that species with intermediate survival ability are more likely to evolve mutualisms than are species with either high or low survival abilities. His arguments are restricted to mutualisms in which two or more individuals live together intimately over long periods of time (symbiosis). These interactions often take the form of a parasite–host interaction, with one species acting essentially as the host for the other species and, indeed, Roughgarden assumes that the mutualisms he considers pass through a parasitic phase.

Roughgarden suggests that if host survival ability is low, then the evolution of a mutualism is unlikely because the probability of encounter between an individual host and a potential mutualist is also usually low Furthermore, the interaction may be too risky for the symbiont even if it finds a host because the host may die. If host survival ability is very low in the absence of the symbiont, then the symbiont must be able to raise host survival considerably to make the interaction beneficial to the symbiont. Together the low probability of encounter and low host survival even after encounter with the potential mutualist make mutualisms unlikely to evolve among species with low survival ability.

At the other extreme, if host survival ability is already high, then the chances for the evolution of a mutualism are again low because the symbiont may have to make a very high input into the interaction in order to increase significantly host survivorship. (The input will usually be in the form of protection or nutrition.) Therefore, it is among symbiotic interactions in which the host exhibits intermediate survival ability that the evolution of mutualism is most likely.

Roughgarden's intermediate-survival-ability model applies primarily to mutualisms in which each individual must encounter a potential mutualist independent of others in the population. If symbionts are passed between host parents and offspring or among siblings, however, then low survivorship of individual hosts becomes less of a restriction on the evolution of symbiotic mutualisms. For example, gut symbionts are passed from parents to offspring (Wilson, 1980) and among siblings (e.g. termites, Noirot and Noirot-Timothea, 1969) in some insect species. Moreover, possible death of the host after discovery by the symbiont is a barrier to the evolution of mutualisms only in the specific kinds of mutualism with which Roughgarden is concerned: that is, symbiotic associations. Therefore, Roughgarden's specific model is most applicable in mutualisms in which symbionts find hosts independently and the host and symbiont interact for major portions of their lives. His arguments are not extended to mutualisms based on short-term interactions such as those between flowers and flower visitors or seeds and vertebrate or ant dispersal agents.

If the restriction of extended symbiosis is removed from the model, however, then the relationship between intermediate survivorship and mutualism also reflects a relationship between intermediate disturbance regimes in communities and mutualisms. Intermediate disturbance regimes are often associated with high levels of species diversity (Connell, 1978; Hubbell, 1979; Huston, 1979; Hartshorn, 1980), which should allow for a high richness of biotic interactions and, as argued in Chapter 4, the richness of mutualistic interactions in communities is dependent upon the richness of antagonistic biotic interactions. Therefore, species with intermediate survival abilities and communities with intermediate disturbance regimes both provide conditions especially favorable for the evolution of mutualisms.

On a broad scale, some of the most obvious mutualisms in communities, including ant–plant and vertebrate–seed mutualisms, are associated with intermediate levels of disturbance. Light gaps in forests induce a high richness of biotic interactions through intense competition among plants and interactions between animals and plants. Herbivory rates on plants adapted specifically to these kinds of disturbance can be high relative to more persistant plants (Coley, 1980), and ant activity is also often high at these sites (Talbot, 1934; Bentley, 1976). In Costa Rican forests extrafloral nectaries and, therefore, probably ant–plant mutualisms are more common along forest edges and in light gaps than in other forest sites (Bentley, 1976). The more specific mutualisms between ants and myrmecophytes (plants inhabited by mutualistic ants) are also most common in disturbed sites, predominantly in second growth wet habitats (Janzen, 1974b). In addition, almost all of these myrmecophytes have their seeds dispersed by mutualistic vertebrates (Janzen, 1969).

Temperate herbs, shrubs, and vines with fleshy fruits tend to predominate in light gaps and forest edges, as do plants with small-seeded fleshy fruits in tropical forests (Morton, 1973), and removal rates of fruits are higher in these habitats than under closed forest canopies (Thompson and Willson, 1978). Furthermore, animal-dispersed plants generally predominate on rotting logs and the pits and mounds created by treefalls, at least in moist deciduous forests of the midwestern United States (Thompson, 1980).

In general, environments characterized by intermediate levels of disturbance and organisms faced with intermediate survival abilities provide focal points for the conditions under which we should expect a high probability for the evolution of mutualisms between species. Under these conditions, a high richness of biotic interactions is maintained within communities and mutualisms are favored through change in outcome of antagonistic interactions to mutualistic interactions because small inputs from a mutualist could affect fitness of individuals significantly.

STRESS AND MUTUALISM

There is, however, another set or axis of conditions that also favors mutualisms. This set of conditions includes organisms with a high probability of encounter and very low premutualism growth rates in environments that impose a high level of physical stress but lack the richness of antagonistic interactions that is the basis for selection in many other mutualisms.

The mutualisms associated with this set of ecological conditions often involve nutrition of a host by a symbiont in nutrient-poor environments. Examples of these mutualistic symbioses (as reviewed by Lewis, 1973) include interactions between coelenterates and dinoflagellates in nutrient-poor tropical oceans (Yonge, 1968); the relationships between algae and fungi that comprise lichens, which are able to colonize sites very poor in nutrients [but see Ahmadjian (1982), who argues that these interactions are antagonistic]; and the tendency of mycorrhizal fungal infections of forest trees to be high under nutrient-poor condition. Janos (1980) notes that obligately mycotrophic plants (i.e. those requiring mycorrhizal formation to reach maturity under the nutrient conditions in their natural habitats) are most likely to develop and dominate on soils low in availability of minerals. Among plants that are facultatively mycotrophic, the general pattern is better growth without mycorrhizae or lack of development of mycorrhizae on fertile soils but better growth with mycorrhizae on poor or declining soils. In these nutrient-poor environments, as in those characterized by intermediate disturbance regimes and species with intermediate survival ability, small inputs by a mutualist can potentially have major effects on fitness.

Perhaps the most graphic illustrations of how environments that impose unusual physical stresses on organisms have favored novel forms of interaction including mutualisms involve the reversal of the usual trophic order of life. Insectivorous plants and ant-fed plants are the two major ways in which plants have evolved to gain nutrients directly and actively from animals, one in an antagonistic and the other in a mutualistic manner. Approximately 535 species of insectivorous plants and 201 species of ant-fed plants have been described, although mostly based on observations rather than on experiments (Thompson, 1981b). These reversed animal–plant interactions are almost all restricted to environments very low in available nutrients (Janzen, 1974c), and they share many ecological similarities (Table 5.1).

The mutualistic subset of these reversed trophic interactions involves ants that live within or under a part of the plant that appears specialized for harboring ants and absorbing nutrients from the ants' debris piles (review in

Table 5.1.
Similarities Between Insectivorous and Ant-Fed Plants

	Insectivorous	Ant-Fed
Substrate nutrient level	Very low	Very low
Primary use of animal food	Prob. nitrogen	Prob. nitrogen
Life history	Perennial (a few annual)	Perennial
Worldwide distribution	Tropical– temperate (a few boreal)	Tropical

Source. Thompson, 1981b.

Huxley, 1980). The potential ability of the plants to absorb nutrients from these debris piles has been demonstrated experimentally in a few species (Janzen, 1974c; Huxley, 1978; Rickson, 1979). The ants associated with these plants all appear to nest obligately in host plants.

Ant-fed plants have evolved independently in different areas of the tropics at least six times, and perhaps as many as 12 times, depending upon how the relationships of genera and families are interpreted (Thompson, 1981b). These plants include both angiosperms and ferns. The multiple number of times in which these interactions have evolved suggests that the geographic restrictions of these mutualisms is not a result solely of a lack of evolutionary opportunity but is instead a result of restrictions imposed by differing environments. Most suggested ant-fed plants are epiphytes of open-canopy trees, although some are epiphytes in more moist environments (Huxley, 1980). These epiphytes of open-canopied trees are in habitats poor in nutrients for the epiphytes, usually in environments in which the atmospheric absorption habit or tank habit of acquiring nutrients, common in bromeliads (Benzing and Renfrow, 1974), is untenable because of insufficient moisture. These habitats are also generally unavailable to the evolution of an insectivorous habit, because most insectivorous plants rely on alluring glands to attract animals and on water-filled pitchers or other glands to break down the insect nutrients. The use of freely secreting and absorbing glands on the plant surface would act in opposition to the need to conserve water in these habitats. Hence ant-fed plants and insectivorous plants are evolved in response to similar ecological problems but in different types of habitat.

Both ant-fed plants and insectivorous plants are restricted to severely nutrient-poor habitats apparently because the effect of animal-based nutrients seems to diminish rapidly as nutrient levels in the substrate increase. The only experimental evidence in support of this hypothesis, however, is for insectivorous plants (Sorenson and Jackson, 1968; Chandler and Anderson, 1976), although the results of these experiments certainly must be applicable to ant-fed plants as well. Therefore, the evolution of insectivorous plants and ant-fed plants is restricted for the most part to environments very low in substrate nutrients apparently because it is only in these environments that a small input from insects can have significant effects on plant survival or growth rates.

It is certainly not true, however, that these nutrient-poor environments are the only conditions that favor mutualisms involving nutrition of a host by a symbiont. Herbivorous mammals in general rely heavily on their gut symbionts for nutrition in all environments (McBee, 1971; Allison and Cook, 1981). Unusually nutrient-poor environments, however, increase the probability of novel mutualisms between species because small inputs of nutrients from a symbiont can increase significantly host growth rates.

Notice the similarity in logic of this stress hypothesis of mutualisms with the logic in the intermediate survival/disturbance model. These two sets of environmental conditions represent the extreme conditions under which small inputs by a symbiont can have the greatest effect on host survival and/or growth rate. The stress hypothesis emphasizes the evolution of mutualisms in environments that impose severe restrictions on growth rates through limitations imposed by the physical environment. The intermediate survival/disturbance model also emphasizes the evolution of mutualisms in environments where the symbiont's potential effect on host growth rate, survival, or reproduction is high, but here through the intensity of biotic interactions with other species. It is in these two kinds of environments that we should search especially for new kinds of mutualism; although of the two sets of conditions, those in the intermediate survival/ disturbance model are more likely to allow the maintenance of the greater richness of mutualisms over evolutionary time because of the relationship between the richness of antagonistic and mutualistic interactions (Chapter 4).

The point of these two sections has been to ferret out several traits of organisms and characteristics of environments that seem to be associated with high probabilities for the origination or maintenance of mutualisms. This is not meant to be an exhaustive set of life history patterns favoring the evolution of mutualisms. Rather, these are hypotheses for the conditions under which the interaction structure of communities should be particularly rich in mutualistic interactions.

SOCIAL BEHAVIOR AND MUTUALISM

Having established some general aspects of life histories and habitats that seem to be associated with high probabilities for the evolution of mutualisms, the next step is to consider specific components of life histories that favor mutualisms. Only one specific aspect of life histories will be developed here, but it is a particularly important focal point for the evolution of mutualisms—the relationship between social behavior and mutualism. This relationship represents an area of study in which evolutionary ecology, population ecology, and behavioral ecology are especially intertwined.

Social Insects and Mutualism

Many of the examples in these chapters on mutualism involve social insects. This is not simply bias or coincidence. The evolution of eusociality in insects has often involved concomitant or subsequent evolution of mutualisms with other organisms for resources critical to colony life. The colony life of eusocial bees is centered on their often mutualistic interactions with flowers. The social framework of termite colonies involves the transfer between individuals of their cellulose-digesting gut symbionts. Ants, especially, are the preeminent mutualists: no other single family of organisms is comprised of species involved in as broad an array of mutualisms as are ants. Ants protect from enemies plants in many angiosperm families (Bentley, 1977b; Tilman, 1978; Inouye and Taylor, 1979; Schemske, 1980), larvae in the butterfly families Lycaenidae and Riodinidae (Hinton, 1951; Ross, 1966; Atsatt, 1981b; Pierce and Mead, 1981), and nymphs and adults of many Homoptera (Way, 1963) in order to milk their hosts of nutritive secretions. Some plants develop hollow chambers in which ants can live safely and the plants are provided in the process with nutrients left in the ants' debris piles, as discussed in the previous section. Many angiosperm seeds are adapted for dispersal by ants, and ants are provided with lipid-rich food in the transaction (Berg, 1958; Culver and Beattie, 1978, 1980; Handel, 1978; Horvitz and Beattie, 1980; Thompson, 1981a).

These interactions between social insects and other organisms have a long history of study by naturalists. Bequaert's (1922) bibliography of over 1100 references concerned at least in part with interactions between ants and plants provides an indication of the long and sustained interest in interactions between social insects and other organisms, even prior to the explosion of studies in the twentieth century. (It is also a humbling experience for anyone trying to assimilate the literature in this field of study.) Nonetheless, the study of the selection pressures that have produced the richness of interactions involving social insects is a small subset of the descriptive analyses of these interactions.

Sociality and mutualism in insects may often reinforce one another over evolutionary time. The evolution of eusociality has been considered generally from the viewpoints of either genetic relatedness among siblings (Hamilton, 1964, 1972; Bartz, 1979), parental manipulation (Alexander, 1974), or defense against predation (Lin and Michener, 1972; Evans, 1977). The frequency with which eusocial insects are involved in mutualisms, however, suggests that these interactions have also been important as selection pressures in the evolution of these insects. Mutualistic interactions may often provide colonies with ready and continual access to resources necessary for the maintenance of a stable colony structure. Pollen and nectar can be stored by bees, aphids can be milked regularly by ants, and cellulose-digesting gut symbionts can make the cellulose-packed world surrounding a termite colony one large food resource. Comparisons of closely related eusocial species that differ in the extent to which they interact mutualistically with other species may provide new insights into the evolution of social structure in these organisms that Darwin (1859) referred to as "the one special difficulty."

The relationship between sociality and mutualism can also be turned about to ask why social insects are involved in such a wide array of mutualisms and with such a broad number of taxa. To focus the problem a bit more, the question really hinges on why ants, with their peculiar life histories, have been so prominent in generating mutualistic interactions in terrestrial communities.

Ants are nearly ubiquitous in terrestrial habitats—they nest everywhere that they can find safe nest sites for the colony from subterranean environments to forest canopies—and they often forage as scavengers for a wide variety of foods. These aspects of their life histories contribute to the likelihood that ants interact worldwide with an unusually large number of other species.

Nonetheless, these traits seem insufficient to account fully for the broad range of mutualisms involving ants. In the variety of mutualisms that have evolved from interactions between ants and other species, the important additional trait of ants is the functioning of an ant colony as a single social unit. A colony of ants can protect a shrub or a small tree from herbivores, whereas a single ant or a pair of ants could not. A colony of ants can protect a small herb or clone of aphids from herbivores because the same ant from the colony need not remain continuously on the plant or with the aphids; at least a few other ants from the colony will also visit the plant or the aphid clone routinely—and these same individuals will often return day after day

(Ebbers and Barrows, 1980). A colony of ants can generate enough excretion and debris to provide a significant input of nutrients into a plant in which the colony is nesting, but a single ant probably could not. (A single large animal nesting in a hollow in a tree, however, could provide a significant input of nutrients to the plant, and Janzen (1976b) considers this as a possible explanation for why the heartwood of some trees rots.) In general, the social structure of ants is critical to the origin and maintenance of many of the mutualisms with which ants are involved.

Furthermore, the extraordinary richness of behaviors associated specifically with social life has often provided the basis for mutualisms involving ants. Wilson (1976a,b; Wilson and Fagen, 1974) categorized the broad range of behaviors exhibited by ants in defense of nests, in food gathering, and in interactions among colony mates; the behavior of ants toward mutualists is derived generally from these social behaviors. Potential mutualists can exploit this rich repertory of ant social behavior through evolutionary changes that tap one subset of ant behaviors while excluding other subsets. In this way the already existing broad range of ant behaviors may facilitate the evolutionary transition between antagonism and mutualism, and may explain partly why these social insects are involved in so many mutualisms.

The evolution of these mutualisms, however, may involve not only the richness of social behavior in ants but also the distribution of those behaviors among individual ants. Wilson (1975) argues that a major reason ants harbor so many kinds of symbionts—not just mutualists—is because individual ant castes do not exhibit the whole range of behaviors common to the colony. Therefore, breaching the social structure of an ant colony may require only that the symbiont mesh with a certain subset of the behaviors of ants in the colony to be considered effectively as part of the colony. Since many of the behaviors of ants in worker castes are directed already toward the feeding and protection of other individuals, the initial transfer of a caste's subset of behaviors to other organisms that can breach the ant's communication system may occur quickly in evolutionary time for any particular interaction. The great degree of morphological specialization found in species that live as symbionts with ants (Wilson, 1971), however, indicates that the behavioral modifications alone often cannot allow for complete assimilation of a symbiont into a colony's social structure, especially if the symbiont is parasitic rather than mutualistic.

For mutualistic interactions, the major point remains that the behavioral richness and complexity of the social life of ants provides the evolutionary basis for the origin and maintenance of these relationships. Consider, for example, the evolution of mutualisms in which ants protect other animals or plants from enemies while the ants obtain food from the mutualist. Ants prey commonly on a wide variety of other insects, but many of these same ant species do not usually kill the insects from which they derive nutrition through glandular secretions. Skinner and Whittaker (1981) demonstrate that foraging by the ant *Formica rufa* in forest canopy in northern England significantly reduced populations of winter moth *(Operophtera brumata)* larvae and aphid species not tended by ants because of predation on these species by the ants. On these same trees, however, populations of aphid species tended by ants for honeydew increased dramatically (800% in one species) over control populations lacking ants. The aphids have transformed their interactions with ants from what must have been initially antagonistic interactions into mutualisms by exploiting a different subset of ant behaviors. The solicitation behavior that ants use to cause aphids to secrete honeydew is essentially the same as the behavior used among colony mates for regurgitation of foods from one ant to another (Wilson, 1971).

Similarly, the evolution of extrafloral nectaries on plants may have involved exploiting these same feeding habits in ants and transferring behaviors associated with colony life or tending of Homoptera directly to the plant itself. By short-circuiting the mutualism between Homoptera and plants through production of extrafloral nectaries, the aggressiveness of the ants can be transferred from defending the homopterans to defending the plants. From the ant's perspective, although honeydew-excreting Homoptera may provide ants with food rich in amino acids and lipids as well as carbohydrates, tending of Homoptera may require more complex care than feeding directly from extrafloral nectaries (Carroll and Janzen, 1973). Alternatively, production of extrafloral nectaries could evolve initially as a defensive response to ants feeding on the growing tips of plants, so that ants tend the growing tips for nectar rather than chew directly into moist and nutritious tissue. Additionally, ants may be less likely to remove plants with extrafloral nectaries from areas surrounding their nests than they are willing to clip other plants.

No matter what the initial selection pressures for these mutualisms, the multitude of times in which these interactions have arisen among families of angiosperms (Zimmerman, 1932; Bentley, 1977a) and even within families (Elias, 1980) suggests the relative ease evolutionarily with which ant behaviors can be channeled into protection of plants through the production of extrafloral nectaries. Moreover, a mutation in a plant causing production of extrafloral nectar can be beneficial instantaneously to that individual plant. Green beans *(Phaseolus vulgaris)* placed in a Costa Rican forest with droplets of sugar water on the leaves suffer lower levels of herbivory through ant activity on the plants than do bean plants lacking the artificial nectar (Bentley, 1976).

Other mutualisms involving ants require many different and separate behaviors on the part of the ants, and it is difficult to imagine the evolution of these mutualisms without prior sophistication in social behavior. Some Homoptera and some butterfly larvae are not only milked for their secretions, but are carried to appropriate places on plants in order to feed (Wilson, 1971; Atsatt, 1981b). The larvae of the Mexican riodinid butterfly *Anatole rossi* are kept by ants *(Camponotus abdominalis)* in pens at the base of the butterfly's host plant during the day. The larvae are shepherded to higher parts of the plant at night to feed and are returned to the pens by dawn. During cool periods in winter, the ants convert the pens at the bases of the larval food plants to tunnels for the larvae (Ross, 1966).

Similarly, ant gardens found in Central and South American seem to represent multifaceted mutualisms in which the ants participate in many aspects of the plants' life histories. The term "ant garden" (Ameisengärten) was coined by Ule (1901) to describe the arboreal carton nests of some neotropical ant species and the characteristic plant species that grow on the outside of these nests. At least some of the angiosperms and ferns associated with ant gardens usually grow only on these carton nests (Ule, 1906; Moore, 1973; Prance, 1973; Macedo and Prance, 1978).

The natural history of these interactions is incompletely known. Some carton nests of ants may be built around the roots of plants established epiphytically on trees (Prance, 1973) or the ants may collect actively the seeds of particular plant species and plant them in their carton nests (Ule, 1901). The only detailed field study on ant garden plants, however, is Kleinfeldt's (1978) work on *Codonanthe crassifolia* is Costa Rica. The ants associated with *Codonanthe* feed on the floral and extrafloral nectar and the fruit pulp and seed arils of the plant. The ants can carry away all of the 100 or more tiny seeds contained in a single red berry within two days (pers. obs.). Kleinfeldt found that newly built nests are not always associated initially with *Codonanthe* plants, indicating that the seeds are often brought to the nests by the ants and germinate in already established nests. Growth rates are higher for *Codonanthe* plants associated with active ant nests than for plants not associated with ant nests.

Therefore, these ant garden mutualisms may involve pollination of the plants by ants visiting the floral nectaries, seed dispersal and planting in a safe site for germination, enhanced growth rates for the plants through the nutrients derived from the carton nest, and protection from herbivores as a result of ants patrolling the plants while they feed on extrafloral nectar. The ants benefit certainly from the readily accessible food in the forms of floral nectar, extrafloral nectar, fruit pulp, and seed arils. Also the roots of these plants may strengthen the ants' carton nests, as suggested by Ule (1901). In a very real way these plants have become an integral component of the social system of these ants.

General Considerations

The evolutionary basis for mutualisms involving ants suggests a more general hypothesis regarding the relationship between sociality and mutualism: within a taxon in which species vary in their degrees of social behavior, mutualisms with other species should be more common among the more social species. These social behaviors include, for example, parental care, territoriality, and group defense against predators. The logic is the same as that used for the commonness of mutualisms involving ants—the richer the existing set of social behaviors, the more flexible and variable are the interactions with other species, and the greater the probability that some antagonistic or commensalistic interactions can be transformed evolutionarily into mutualistic interactions.

There is, however, an immediate problem in evaluating the hypothesis. Neither social behavior nor mutualism is actually an independent variable. Gilbert (1975) argues that animal–plant mutualisms depend upon the behavioral sophistication of the animal but these interactions also partly determine the animal's behavioral sophistication. Hence just as the richness of social behaviors may increase the probability of the evolution of mutualisms with other species, so may some mutualisms allow for the evolution of new social behaviors.

These new social behaviors can arise potentially through the increase in time available to organisms whose foods are readily accessible because of mutualisms. The extraordinary behavioral sophistication of *Heliconius* butterflies in Central America is a result partly of their mutualism with *Anguria* flowers (Gilbert, 1975; 1977). *Heliconicus* butterflies have evolved the ability to utilize pollen as adults (Gilbert, 1972; Dunlap-Pianka et al., 1977), and this has allowed adults to live longer than most butterflies as they search scattered *Passiflora* hosts on which to lay eggs. *Heliconius* butterflies trapline their *Anguria* flowers daily and roost gregariously at night.

Similarly, D. W. and B. K. Snow argue that the evolution of lek behavior in small altricial birds is often associated with frugivory because frugivorous birds are able to satisfy their food requirements in a small portion of the day (Snow, 1971; Snow and Snow, 1979). The food of these birds is advertised readily by the plants, so the time spent searching for food is decreased relative to fully insectivorous birds. Not only can adults quickly satisfy their food requirements, but in some species frugivory contributes to the release of the male from nest-tending activities and these males spend a great amount of time in competition for mates (Snow and Snow, 1979; Snow, 1980). Hence some of the most sophisticated courtship behaviors in small altricial birds occur among the manakins and bellbirds, which feed entirely or largely on fruits, the extent of frugivory varying with species (Skutch, 1949; Snow, 1962a,b; Sick, 1967; Snow, 1970; Schwartz and Snow, 1978). Similarly, Ochre-bellied Flycatchers *(Pipromorpha oleaginea)*, which also feed to a great extent on fruits, exhibit a mating system in which males court in loose leks (Skutch, 1960; Snow and Snow, 1979).

In general, the relationship between the evolution of complex social behaviors and mutualisms seems to be a worthwhile direction in which to search for components of life histories that increase the probability that mutualisms will evolve between species. In addition, attention to the kinds of mutualisms in which species are involved may reveal additional patterns in the evolution of social behavior within phylogenetic lineages.

CONCLUSIONS

The major points of this chapter all concern components of the life histories of organisms in environments that seem to be foci for the evolution of mutualisms:

1.   Organisms living in environments characterized by intermediate levels of disturbance and faced with intermediate survival abilities have a higher probability of evolving mutualisms with other organisms than organisms with either very high or very low survival abilities in similar environments rich in biotic interactions. Under such conditions, small positive effects of a mutualist can increase survival or growth rates significantly.

2.   Organisms living in environments characterized by low richness of antagonistic interactions but high levels of physical stress (e.g. nutrient-poor habitats) have a high probability of mutualistic encounters with other species because small input by a potential mutualist can affect growth rates significantly.

3.   The evolution of mutualisms is associated positively with the richness of social behavior in species such that (a) mutualisms should be more common among the more social species within a taxon and (b) the richness of social behaviors within a species may be partly an evolutionary result of mutualisms that allow species to spend less time foraging for food.

These points are all hypotheses regarding how the interaction structure of communities should be expected to vary as life history traits, physical conditions, and the richness of biotic interactions vary. As in antagonistic trophic interactions and in competition, if we are able to identify the kinds of ecological conditions that select for the evolution of mutualistic interactions, then we will be able to assess better how selection acts over evolutionary time to change the outcomes of interactions between antagonism and mutualism. In the absence of any such attempts, the ecological study of interactions will remain a series of "neat stories," of which the only purposes of each study are to determine (1) whether a particular interaction is either antagonistic, commensalistic, or mutualistic, and (2) how much the fitness of each participant is affected by the interaction.

CHAPTER

6

EVOLUTION OF
MUTUAL DEPENDENCE

This chapter is intended as an exploration into the range of selection pressures affecting specificity among mutualists parallel to the comments in Chapter 2 on the selection pressures affecting specificity in parasites, grazers, and predators in antagonistic interactions. Primarily, I consider similarities and differences in the ways selection acts on the evolution of specificity in mutualisms as compared with antagonistic interactions, and on visitors as compared with hosts. The tentative tone of the chapter emphasizes how little we really understand the range of selection pressures that affects the degree of mutual dependence between mutualists.

VISITORS AND HOSTS

In most mutualisms one species is identifiable as a host for another species, the host providing either food or a safe place to live for the visitor. Plants are hosts to animals that pollinate their ovules, disperse their seeds, protect their leaves, or fertilize their roots. Corals are hosts to shrimp and crabs that protect them from grazing starfish (Glynn, 1980), and oropendulas are hosts sometimes to cowbird nestlings that kill the botflies on their oropendula nestmates (Smith, 1979). Selection on hosts in mutualisms differs from selection on visitors, so I consider hosts and visitors in separate sections of this chapter.

Separation of mutualistic pairs into visitors and hosts creates some semantic problems when considering some particular mutualisms, but these problems are inevitable in any attempt to categorize interactions between organisms. The purpose here is not just to categorize but rather to show how and why selection can differ on mutualists in the same interaction. As an example of a semantic problem, consider dispersal of seeds with fleshy fruits by birds as compared with dispersal of seeds with elaiosomes by ants. Fleshy fruits are usually held on the plant until eaten by a bird. The plant makes many fruits and acts as the host to the birds that visit the plant. Plants with elaiosomes, however, generally drop their seeds to the ground and the ants interact only with the seeds, not the parent plant. It would be difficult to think of the seed being carried by the ant as the host to the ant, and it would be improper to do so. The host is still the parent plant that made the seed and packaged it with an elaiosome. It is the parent plant whose investment in the mutualism is subject to selection by the ants, although the interaction between the host parent plant and the visitor ant occurs only indirectly through the seed laying on the ground. In the following sections, then, species are designated as hosts or visitors not for any typological reasons but simply as an aid to understanding how selection acts in the evolution of mutual dependence between species.

VISITORS

Selection seems to act on visitors in mutualisms in two opposing ways. First, to the extent that visitors in mutualisms have behaviors similar to parasites, grazers, or predators, selection on specificity acts in the same manner as in antagonistic interactions. This is not surprising and is not simply analogy since many mutualisms derive from antagonistic interactions. Therefore, parasitelike mutualists seem to be more often restricted to single hosts than grazer or predatorlike mutualists, whose use of hosts is often plastic and governed by learning.

Opposing this axis of selection is a second kind of selection generally absent from antagonistic interactions and originating in selection on the host. Over evolutionary time hosts in mutualisms are under selection to increase rather than to decrease their availability to visitors. This selection on hosts has two potential effects on selection on visitors; both effects mitigate the patterns of specificity on hosts expected if selection acted only the same as in antagonistic interactions. First, selection on parasitelike mutualists to be specific to one host species may be less stringent than on true parasites, since selection on the visitor need not be focused on overcoming the host's defenses. As a result, parasitelike mutualists may sometimes have broader ranges than comparable true parasites. Second, grazer or predatorlike hosts may be able to specialize on a single host species more easily than true grazers or predators, since selection on hosts in at least some types of mutualism is toward increasing visits by these visitors. Therefore, selection on hosts in some mutualisms can be toward providing resources that fulfill the dietary or other requirements of the visitor, thereby allowing visitors to restrict their visits to a single host species.

Ecological patterns in specificity on hosts, then, reflect the evolutionary balance between these opposing selection pressures (plus other selection pressures acting on hosts considered in the next section). The way selection balances these conflicting pressures must vary with (1) the type of mutualism, (2) the overall life histories of the mutualists, and (3) the community context of the interaction. Each of these is considered in this section. Understanding that selection on visitors in mutualisms is only partly the same as selection on parasites, grazers, and predators at least prevents knee-jerk extensions of the logic of host selection and specificity in antagonistic interactions to mutualisms.

Specificity in Pollinators

The behaviors of pollinators show how selection pressures are balanced in different ways depending upon the evolutionary origin of the interaction and the life histories of the visitors. Most mutualisms between flowers and pollinators involve flower visitors that essentially graze on the nectar or pollen of many flowers during their lifetimes. A few pollinators, however, live as parasitelike mutualists on their host plants. The classic examples are the agaonid wasps that pollinate figs and the *Tegeticula* moths that pollinate yuccas. Both of these mutualisms derive evolutionarily from antagonistic interactions in which the larval wasps and moths developed as parasites in the floral parts of the hosts (Chapter 4). Of the 900 or so species of *Ficus*, each seems to have its own host-specific wasp pollinator in most cases (Ramírez, 1970; Janzen, 1979b; Wiebes, 1979). The level of host specificity is more difficult to evaluate in the yucca moths. The taxonomy of the 40 or so species of *Yucca* is not finalized (Davis, 1967) and the four recognized species of *Tegeticula* may include a number of hidden species currently lumped together in analyses based on morphology (Davis, 1967). In any case, these yuccas and moths are intertwined much more closely than the vast majority of interactions between plants and their pollinators.

The extreme mutual dependence between agaonids and figs and between yucca moths and yuccas seems to be simply an extension of selection for extreme host specificity common in parasites, and especially in parasites that live internally on their hosts. These interactions seem unusual because, in general, the habits of grazers are much more likely to result in effective pollen transfer than the habits of parasites. In these few instances the oviposition behaviors of the progenitor wasps and moths selected for the evolution of pollen transfer by the parasites (Chapter 4). The ecological result reflects selection on the visitor the same as that on true parasites without any indication of counterselection to broaden the host range.

Selection on grazerlike pollinators, by comparison, seems more of a balance of opposing selection pressures than selection in the unusual instances of pollination by parasitelike mutualists. Many solitary bees exhibit an extreme level of specificity on host plants (oligolecty) that is rare among antagonistic grazers of plants. Therefore, selection is not simply an extension of the selection pressures that usually prevent specialization on single host species among antagonistic grazers. Oligolectic bees have a host species or small group of species to which they restrict their pollen-collecting visits even when other pollen sources are available (Linsley and MacSwain, 1957), although during times of shortage of their host plant or when foraging for nectar, they visit other plant species (Thorp, 1969; Eickwort and Ginsberg, 1980). The seasonal flight period of these bees is restricted largely to the blooming times of their hosts (Linsley, 1958).

The pollen and nectar of the host plants of oligolectic bees is a predictably available food of high nutritional value and selection on the host must be toward remaining predictably available to the pollinator. Plants flowering outside the times that other individuals in the population are in flower may receive fewer visits by pollinators or at least fewer visits by pollinators carrying conspecific pollen (Augsberger, 1981). Under these ecological conditions, restriction to a single plant species is possible for grazerlike pollinators that can evolve a life history with a phenology that takes advantage of these boom and bust periods of food availability.

No absolute statements, however, can be made about the degree of host specificity among oligolectic solitary bees. Within oligolectic bee taxa the degree of host specificity varies: some bees seem restricted to a single plant species, others visit several species within a plant genus, and still others visit several genera within a plant family. Some of this variation certainly reflects differences in host speciation as compared with bee speciation. A bee species seems less specific if its host plant is undergoing speciation, since the bee is now associated with several species in the plant genus. In general, most oligolectic bees are restricted to a plant genus throughout their range and often at the local population level to particular plant species within the genus.

Closely related species of bees, however, may be restricted to very different plant genera. The bee genus *Andrena* subgenus *Diandrena* is comprised of 25 species that occur in western North America from the eastern slopes of the Rocky Mountains to the Pacific coast. Most species are restricted to California. All the bees in this subgenus are oligolectic, but the species differ widely in the host plant genera and species to which they are restricted (Thorp, 1969; Table 6.1).

The restriction of these grazerlike pollinators to particular plant genera or species, however, does not indicate necessarily that the plants and the particular pollinators are completely mutually dependent. The plants may be hosts to several oligolectic bee species, so the pollinators are more specific than the plants. The *Diandrena* bees (Table 6.1), for example, are not the only oligolectic pollinators associated with *Camissonia* and *Clarkia* in the Onagraceae. A total of 18 species of *Andrena* bees are known to be oligolectic on *Camissonia* species, nine bee species in the subgenus *Diandrena,* and nine in the subgenus *Onagrandrena* (Linsley et al., 1973). Ten species of oligolectic bees are known from species of *Clarkia* (MacSwain et al., 1973).

A tendency toward oligolecty is not a universal trait of grazerlike flower visitors. Some life histories prevent the opportunity for extreme specificity on flowers. Social bees must continually replenish pollen and nectar stores and genetically based oligolecty does not occur generally among these species (Linsley, 1958; Eickwort and Ginzberg, 1980). Instead, individual grazers within colonies learn how to handle different

Table 6.1.
Distribution of Food Plant Genera Among the 25 Species of the Bee Genus *Andrena* Subgenus *Diandrena* in Western North America

Plant Genus	Plant Family	Number of Bee Species
Camissonia	Onagraceae	9
Clarkia	Onagraceae	1
Ranunculus	Ranunculaceae	1
Arenaria	Caryophyllaceae	1
Lasthenia	Compositae	2
Blennosperma	Compositae	1
Malacothrix	Compositae	1
Anisocoma and *Malacothrix*	Compositae	1
Agoseris	Compositae	2
Agoseris and *Microseris*	Compositae	1
—	Compositae (ligulate)	5

Source. Data from Thorp (1969).

plant species and remain constant on these flowers until floral availability changes (Heinrich, 1979). Similarly, some hummingbirds are constant to one plant species at least for several days at a time (Linhart and Feinsinger, 1980) but must change hosts as floral availability changes. Unlike solitary bees, neither social bees nor nectarivorous birds generally have life histories that allow selection for genetically based dependence on a single plant species.

In summary, parasitelike pollinators seem to retain the level of host specificity common to antagonistic parasites with little tendency to broaden their host range. In contrast, some but not all, grazerlike pollinators are able to be more host specific than many antagonistic grazers.

Specificity in Frugivores

Frugivores do not encompass the wide range of specificity found in flower visitors. All mutualisms between plants with fleshy fruits and fruit-eating animals involve animals whose foraging behaviors are derived from progenitors that fed as predators or grazers on plants or animals. For most frugivores, fruit is still only a part of their diet. Foraging patterns and specificity in frugivores is affected significantly by the patterns of selection on hosts and these selection pressures on hosts are considered in the next section. It is worthwhile here, however, to consider from the frugivore viewpoint how selection acts to influence the degree of host specificity.

Plants produce two kinds of fleshy structures that are attached to their seeds and made for consumption by animals. Elaiosomes—usually lipid-filled gelatinous bodies attached to the seed coat—are made especially by many temperate forest herbs (e.g. Beattie and Culver, 1981), plants in the arid regions of Australia (Berg, 1975; Davidson and Morton, 1981), and some plants in moist tropical forests (Horvitz and Beattie, 1980). None of these mutualisms is known to involve ants that are specific to the elaiosomes of any one plant species or to involve plants that rely on one ant species for seed dispersal. The elaiosomes are simply one food item in a wide array eaten by ants. The food resource is ephemeral relative to the length of life of an ant colony, so as in social bees that visit many flower species, the social life of ants does not permit restriction to a single plant species for elaiosomes.

Most other plants that produce a fleshy covering over their seeds rely on vertebrates to disperse their seeds. Usually, fruit is only a part of the diet of vertebrate frugivores, and few vertebrates rely strictly on fleshy fruits to feed both themselves and their young. Fruit may rarely be favored as the total diet for nestling birds because either fruits have a low protein content (Morton, 1973) or because they have a low protein/calorie ratio (Ricklefs, 1974; Foster, 1978). Either situation could result in slower growth rates than would occur on animal diets. A longer nestling period increases the probability that the nest will be discovered by predators, a major source of nestling mortality (Ricklefs, 1969), before the nestlings fledge. The Oilbird and the Bearded Bellbird, both totally frugivorous, do have longer growth rates than more insectivorous species (Snow, 1962c; Snow, 1970; Ricklefs, 1976). Morton (1973) suggests that total reliance on fruits to feed nestlings, therefore, is coupled with selection for use of relatively safe, although probably rare, or very inconspicuous nest sites. The Oilbird nests in dark caves (Snow, 1962c) and the Bearded Bellbird builds very light extremely inconspicuous nests in branches (Snow, 1970). In addition, the Bearded Bellbird lays only one egg and has long incubation stints that decrease movement around the nest; the down of the nestling is very cryptic and the nestling is silent during feeding sessions. The evolutionary tradeoffs for the Oilbird and the Bearded Bellbird, however, are different. The Oilbird has very safe nesting sites and the slow growth of the nestlings allows for a relatively large clutch size with a mean of 2.7. The nesting sites, however, are very rare. In contrast, the Bearded Bellbird relies on inconspicuousness and it has a very low clutch size of one.

The diets of even these obligately frugivorous species, however, are not restricted to single plant species (Table 6.2). Of course, all fruits are not eaten to the same extent: in Snow's (1962c) study of the Oilbird, over 99% of the fruits in the diet were from only three plant families (Palmae, Lauraceae, and Burseraceae) and 40% were from one palm species (*Euterpe langloissi*). Similar variations in the extent to which fruit species are eaten are found in most of the species listed in Table 6.2. Nevertheless, the point remains that no frugivores are known to be restricted to one plant species. Most other avian frugivores, including thrushes, mimids, tanagers, manakins, and some vireos and warblers, eat a combination of insects and a variety of species of fruits (Table 6.2; references in Thompson and Willson, 1979).

In general, the evolution of fruits has not involved selection for species of frugivores that are highly specific to single host species. Since seed dispersal mutualisms have never evolved from parasite–host interactions, the specificity that is a usual part of these interactions has never been carried over into a seed-dispersal mutualism.

Rather, the behavior of frugivores is more like grazers in antagonistic interactions. Nothing in the usual behavior of grazers is disadvantageous in these mutualisms. Therefore, there is probably little selection on the

Table 6.2.
Plant Families and Species Included in the Diets of Some Birds That Feed Largely on Fruits[a]

Species	Study Area	Collection Method	N^b	Number Identified Plant Families	Number Separable Plant Species	Reference
STEATORNITHIDAE						
Steatornis carpensis (Oilbird)	Trinidad	Regurgitated seeds	112.717	10	36	Snow (1962[c])
TROGONIDAE						
Pharomachrus mocinno (Resplendent Quetzal)	Costa Rica	Observation	—[c]	17	43	Wheelwright (pers. comm.)
COLUMBIDAE						
Ducula spirolorrhoa (Torresian Imperial Pigeon)	New Guinea	Dissected crops	40	12	—	Frith et al.[d] (1976)
Ptilinopus coronulatus (Little Coroneted Fruit-dove)	New Guinea	Dissected crops	39	14	—	Frith et al. (1976)
P. iozonus (Orange-bellied Fruit-dove)	New Guinea	Dissected crops	151	13	—	Frith et al. (1976)
P. magnificus (Wompoo Fruit-dove)	New Guinea	Dissected crops	118	17	—	Frith et al. (1976)
	N. Queensland	Observation	—	25	50	Crome (unpubl.)[e]
P. perlatus (Pink-spotted Fruit-dove)	New Guinea	Dissected crops	62	8	—	Frith et al. (1976)

Table 6.2.
Continued

Species	Study Area	Collection Method	N^b	Diet Composition			Reference
				Number Identified Plant Families	Number Separable Plant Species		
P. regina (Red-crowned Fruit-dove)	N. Queensland	Observation	—	8	12		Crome (unpubl.)
P. superbus (Purple-crowned Fruit-dove)	New Guinea	Dissected crops	16	7	—		Frith et al. (1976)
Macropygia amboinensis (Brown Pigeon)	N. Queensland	Observation	—	15	22		Crome (unpubl.)
COTINGIDAE							
Procnias averano (Bearded Bellbird)	Trinidad	Regurgitated seeds	1,833	20	39		Snow (1970)
PIPRIDAE							
Manacus manacus (Black and White Manakin)	Trinidad	Observation, etc.[f]	—	27	105		Snow (1962a)
Pipra erythrocephala (Golden-headed Manakin)	Trinidad	Observation	403	18	43		Snow (1962b)

THRAUPIDAE

Tangara guttata (Speckled Tanager)	Trinidad	Observation	75	7	14	Snow and Snow (1971)[g]
Tangara gyrola (Bay-headed Tanager)	Trinidad	Observation	405	18	33	Snow and Snow (1971)
Tangara mexicana (Turquoise Tanager)	Trinidad	Observation	237	15	26	Snow and Snow (1971)
Thraupis episcopus (Blue-gray Tanager)	Trinidad	Observation	144	15	23	Snow and Snow (1971)
Euphonia violacea (Violaceous Euphonia)	Trinidad	Observation	206	13	19	Snow and Snow (1971)
Chlorophanes spiza (Green Honeycreeper)	Trinidad	Observation	165	11	22	Snow and Snow (1971)

[a]Values are composites for avian species rather than the number of fruit types eaten by individuals, since individual diets are not available in the literature.

[b]Number of dissected crops, observations, or fruits as estimated from regurgitated seeds. *N* not obtainable in this format for all of the studies.

[c]Based on 2000 hours of field observation.

[d]Only species with > 15 dissected crops included from this study.

[e]General summary of feeding habits published in Crome (1975) for all species listed in Table as Crome (unpubl.).

[f]Observations, collections of regurgitated seeds at display grounds and at nests, and a few trapped birds.

[g]Only species in which > 50% of the feeding observations were on fruit are included here.

host to provide resources to the frugivores in such a way that would allow the frugivores to be specific to one plant species. Unlike in pollination, constancy and specificity in the frugivore is not important for the interaction to be mutualistic. In fact, it would probably be disadvantageous to individual plants if all their seeds were deposited under conspecifics by a highly constant dispersal agent that moved only from plant to conspecific plant. Therefore, the evolution of mutual dependence in mutualisms between fruits and frugivores is probably under less selection toward a high degree of specificity than mutualisms between plants and pollinators.

Recently, Wheelwright and Orians (1982) argue similarly that mutualisms involving seed dispersal by animals should be expected to be less specific than those involving pollen dispersal because selection on neither the plant nor the animal favors specificity in seed-dispersal mutualisms. They emphasize that our thinking on mutualisms involving seed dispersal has been influenced too much by studies on plant–pollinator mutualisms.

Selection in the Community Context

Visitors in other mutualisms seem to follow much the same pattern. To the extent that a mutualism relies on the continual presence of the visitor, the selection pressures on the visitor approach those acting on parasites in antagonistic interactions. The stronger the selection on the host to provide for the total dietary or other requirements of the visitor, the greater the selection on the visitor to become specific to that host even if the life histories of the visitors are more like grazers than parasites. The more specific the actions demanded of the visitor or the host for the interaction to be mutualistic, the greater the selection for increased specificity. These selection pressures are mitigated when visitors have life histories that cannot be constrained completely by the phenology of their hosts as, for example, in plant species that bloom for only a short period of the year whereas the life histories of hummingbirds demand that they search for food all year. The tipping of the balance of these selection pressures toward greater or less specificity of visitors for hosts depends partly upon the range of hosts available.

In some mutualisms different species of hosts offer many of the same resources, and related species of hosts especially may differ only slightly in the resources they offer visitors. These differences between hosts may be sufficient to select for visitors that prefer one host to another; year to year fluctuations in populations of hosts and in their phenology may alter the selective value gained from preferring one host over another. That is, specificity of visitors for hosts is a function not only of the type of mutualism and of how the life histories of the visitors and host mesh but also of the community context of the interaction.

In the Elat Nature Reserve in the northern Red Sea, six species of gobiid fish and seven species of burrowing alpheid shrimp interact mutualistically. The goby finds shelter in the burrow made by the shrimp and the shrimp obtains a tactile alarm system, thereby compensating for its poor vision (Karplus et al., 1981; Karplus, 1981). Each goby species is associated with one to seven shrimp species. The community context of these interactions also varies. Five species of gobiids live in the shallow lagoon but only two are found regularly in deep water. Four shrimp species live in the shallow water and three other species live in deep water. In addition, the shallow water shrimp differ in the reefs and the type of sediment in which they construct their burrows, but the deep water shrimp show no separation in habitat.

Using Pielou's (1969) test for association of pairs of species, Karplus et al. (1981) tested the null hypothesis that the occurrence of gobiid and shrimp species in the same burrow is independent for over 750 associations among the species that overlapped vertically. They rejected the null hypothesis for the associations in shallow water but not for the associations in deep water. The associations in shallow water are very specific, whereas in deeper water the gobiids are not associated with any single shrimp species (Figure 6.1). The two gobiid species that are relatively abundant in deep water but not associated with any particular shrimp species (*Amblyeleotris steinitzi* and *Ctenogobiops maculosus*) are also abundant in shallow water where they are very specific to the shrimp hosts with which they are associated. *Lotilia graciliosa* is highly specific to its host in shallow water and also occurs in deep water, but was not found associated with shrimp in deep water. The two *Cryptocentrus* species occur only in shallow water and are also specific to their hosts. The sixth gobiid species was too rare for analysis. Karplus et al. suggest that the habitat separation of the shrimp species in the shallow water, which does not occur in deep water, may account for some of the difference in specificity of the association with depth, but they also note that gobies are less specific in their use of habitats than are the shrimp.

The specific ecological bases for the differences in specificity in these interactions still need to be explored, but the empirical results suggest convincingly that the specificity of these associations varies with community context. If the study had been restricted to a single habitat at a given depth or to either shallow or deep water, the conclusions would have reflected one or the other extreme of specificity, suggesting either that gobiids are very specific or that they are catholic in their association with shrimp.

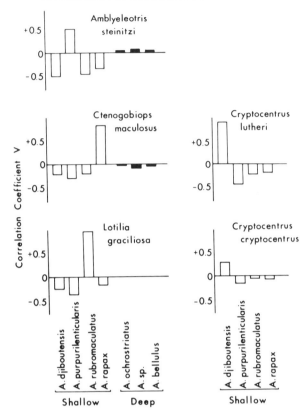

FIGURE 6.1. Specificity of the association of gobiid fish and shrimp of the genus *Alpheus* in the Elat Nature Reserve in the northern Red Sea. Associations in shallow water are given as open bars and associations in deep water as closed bars. The correlation coefficient V (Pielou, 1969) indicates the strength (specificity) of the association. Data from Karplus et al. (1981).

HOSTS

The evolutionary problem for hosts in mutualisms is to simultaneously attract and repel visitors. Selection must act to create a sieve through which only mutualistic visitors with the appropriate sizes, shapes, or behaviors can pass and interact with hosts while excluding other potential antagonistic visitors from the resources used to attract the mutualists. If the selective sieve is too restrictive, potential mutualists are lost and those host individuals are selected against, but if the sieve is too unrestrictive, then other nonmutualistic species gain access to the resources.

The evolutionary problem, then, subdivides into the following components of selection on hosts: (1) selection for a specific subset of visitors, to eliminate thieves and cheaters, and (2) selection for a diversity of mutualistic visitors, to maximize the probability of visits by mutualists. [These two aspects of selection of hosts are old fare among evolutionary ecologists. In *The Naturalist in Nicaragua,* Thomas Belt (1874) in a statement directed clearly at A. R. Wallace's puzzlement over the length of the corolla of the Christmas-star orchid *(Angraecum sesquipedale)* in Madagascar, suggests that naturalists should consider closely not only the adaptations of plants for attracting their pollinators but also the means by which plants prevent nonpollinators from getting nectar from their flowers, if we are to understand floral structure.] It is the simultaneous action of these selection pressures to eliminate thieves but maximize encounters with mutualists that makes selection on hosts in mutualisms different from selection on hosts in antagonistic interactions and on visitors in mutualisms.

Thieves

Thieves or cheaters are an inevitable part of almost any mutualistic system. In pollination ecology the diversity of ways in which some floral visitors gain nectar or pollen without pollinating the plants has even given rise to a specialized terminology: nectar robbers referring to visitors that force entry into the flowers by cutting a hole in the flower and bypassing the opening used by pollinators, nectar thieves referring to visitors that use the usual floral opening but do not collect or deposit pollen while obtaining nectar, and a variety of secondary terms to describe other ways of gaining floral resources without pollinating plants (Inouye, 1980). Selection on hosts in mutualisms to eliminate nonmutualistic visitors can be at best only partly effective. Considering how difficult it is for organisms to eliminate their interactions with all potential parasites, grazers, and predators, it must be even more difficult to offer enticements to one group of organisms to feed on the tissues of a plant or animal without potentially opening the way to other species. In the following discussion I will use terms such as thief and cheater interchangeably and simply as a shorthand for a species that exploits resources generated over evolutionary time through a mutualism but is not itself a mutualist.

Among the species of *Pseudomyrmex* ants that live in the thorns of acacias and protect their host plants is an ant species, *Pseudomyrmex nigropilosa,* that uses the plants without providing protection to its host (Janzen, 1975a). As in other *Pseudomyrmex,* this species is obligate on acacias, nesting in the thorns and feeding on the food bodies. It colonizes acacias that are seedlings or large individuals that have been abandoned by the mutualistic *Pseudomyrmex* and lives in these plants until displaced by a mutualistic *Pseudomyrmex* species or until the plant dies from herbivory or shading. For the acacias to eliminate the thieves from these interactions would probably also entail eliminating the mutualists as well.

Accumulation of thieves may, in fact, be an important selection pressure in the breakdown of mutualisms in general over evolutionary time. From the host viewpoint, mutualisms may break down for the following reasons: (1) the antagonistic species or stressful environments that selected initially for the mutualism no longer affect the host significantly, (2) changes in the demography of the host or the visitor make the probability of encounter between the host and the visitor very low, or (3) species of thieves accumulate on a host over evolutionary time so that the maintenance of resources for mutualists generates more of a cost in fitness than a benefit.

But how much thievery is too much, thereby selecting against production of resources for mutualists? This can be answered only by studying particular mutualisms over time and over the geographic range of hosts. Unfortunately, there are few data even for single seasons of the amount of thievery on hosts in single populations. Most of the data are recent and concern interactions between pollinators and plants or frugivores and plants. For example, Willson and Bertin (1979) determined for *Asclepias syriaca* in Illinois that between 68 and 87% of visits by native flower visitors involved neither the bringing in nor the carrying away of pollen.

The production of elaborate structures to eliminate cheaters may often eliminate only one subset of cheaters and select for another subset. Hummingbird-pollinated plants with long corollas are often robbed of their nectar by visitors that cut holes at the base of their flowers because the nectar is less available directly through the corolla tube. In secondary forest and pasture in the Osa Peninsula of Costa Rica, two hummingbird-pollinated plants *(Aphelandra golfodulcensis* and *Justica aurea)* have at times 90% of their flowers pierced by a robber by noon (McDade and Kinsman, 1980). Certainly, these particular habitats may generate especially high levels of cheaters. Habitat structure, however, changes constantly over evolutionary time, and cheaters may be important selection pressures in the evolution or dissolution of mutualisms at unpredictable intervals of time.

Not all visitors, of course, are absolutely cheaters or mutualists, since the outcome of interactions often depends upon the community context. Where some of the most revealing work can be done on the effects of cheaters on hosts in mutualisms is in interactions in which the same visitor is a mutualist in some populations and a thief in other populations. Howe's intensive studies of seed dispersal in trees with fleshy fruits in

Costa Rica and Panama are among the few that show how the same visitor may be thief, major mutualist, or minor mutualist in different populations from the host viewpoint. The masked tityra *(Tityra semifasciata)* is a common visitor at fruiting trees in Central America. In the wet forest of eastern Costa Rica the masked *Tityra* is a major dispersal agent of the overstory tree *Casearia corymbosa* (Howe, 1977). This same plant species occurs also in the dry forests of northwestern Costa Rica as an understory shrub or small tree, and the masked tityra here is a minor dispersal agent removing only 1.8% of the seeds; the yellow-green Vireo *(Vireo flavoviridis)* removes 65% of the seeds at this study site (Howe and Vande Kerckove, 1979). In Panama, the masked tityra is one of three major dispersal agents of *Virola sebifera* (Howe, 1981) but is a thief of the larger-seeded *Virola surinamensis,* eating the arils without consuming the seeds (Howe and Vande Kerckhove, 1980).

The effect of the masked tityra on dispersal in each of these plant populations and species was determined for a single field season and the actual importance of the masked tityra as mutualist or cheater may vary from year to year. Many other species of birds, and in some cases mammals, also feed on these fruits. Nonetheless, the results suggest how fine the boundary may be between the thief and mutualist. Consistent selection to eliminate all thieves and maximize visits only by certain mutualists may be impossible in many mutualisms.

Since the rigorous study of the role of thieves in the evolution of mutualisms is so recent, it is difficult to suggest any specific patterns in thievery among mutualisms. Some of the questions, however, are these:

1. Should the evolution of cheaters be more common in mutualisms that are more like the interactions between parasites and hosts, grazers and hosts, or predators and prey?
2. What are the patterns of host specificity in thieves as compared with true mutualists?
3. How does the "tolerance level" of hosts for thieves vary among mutualisms?

Maximizing Visits

The elimination of thieves is only part of the problem for hosts. The other part of the problem is which visitors should be favored by selection and how specific selection should be. Perhaps the most general pattern follows again from the differences between mutualisms that are like interactions between parasites and hosts as compared with those like grazers and hosts or predators and prey. Selection on hosts in mutualisms entailing intimate association is often to restrict resources to specific visitors: the entrance chamber (ostiole) into the flowers of figs is usually specific to the appropriate agaonid pollinator (Ramírez, 1974); the morphology and behavior of some burying beetles are adapted for attachment of their specific mutualistic mites (D. S. Wilson, pers. comm.).

Selection on hosts in many other mutualisms, however, seems to be toward maximizing visits by a diversity of potential mutualists with no tendency toward restriction to any one species of visitor. This seems to hold particularly in mutualisms with grazerlike visitors. In a detailed study of the association of ants with the extrafloral nectaries of four *Costus* species in Panama, Schemske (1982) finds no evidence suggesting that *Costus* species have evolved traits that favor visits by any particular ant species. A total of 34 ant species in five subfamilies visit these plants. The *Costus* species differ in some of their associated ants, but this results from differences in the spatial distribution of the plants and the foraging-height preferences, distribution, and abundance of the ant species. Although some ant species seem better at defending the plants than other ant species (Schemske, 1980), the high variation in the likelihood of visitation by any particular ant species probably maintains selection for extrafloral nectar accessible to a wide array of ant species (Schemske, 1982).

Similarly, D. W. Snow (1971) suggests that selection on fruiting patterns in European plants with fleshy fruits has favored plants that are not only conspicuous but also accessible to as many species of birds as possible. This pattern seems to hold as well among North American plants with fleshy fruits. Most forest fruits are eaten by a number of potential dispersal agents and only a few frugivores are excluded from some fruits. For example, red-eyed vireos *(Vireo olivaceus),* which eat many kinds of fruits in late summer and fall, have a difficult time handling the large-seeded fruits of wild black cherry *Prunus serotina* (Willson and Thompson, unpublished data). In the mid-latitude forests of the eastern United States most plants with fleshy fruits ripen their fruits in late August and early September (Figure 6.2; Thompson and Willson, 1979; Stiles, 1980). This timing corresponds closely to the time of maximum numbers of frugivorous bird species migrating south through this latitude. The southward migration of robins together with fewer numbers of hermit thrushes in October creates a second, and sometimes even higher, peak, but this peak is probably exaggerated because of relative increases in robin abundance as compared with other frugivores in recent decades (Thompson and Willson, 1979).

The type of specialization that seems apparent in these temperate

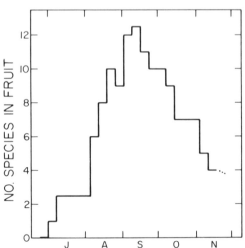

FIGURE 6.2. Number of plant species with ripe fleshy fruits found along a regular census route in Trelease Woods, east-central Illinois throughout late summer and fall. Data from Thompson and Willson (1979).

fleshy-fruited plants, then, is phenological, focused on maximizing visits by frugivorous birds while minimizing the possibility that fruits will be destroyed by microorganisms (Janzen, 1977b), insects (Thompson and Willson, 1978), or mice (Stiles, 1980) before the birds find the fruits rather than on favoring one specific dispersal agent over another. In late August the probability that some fruits in an infructescence will be attacked by insects within a week of ripening is 60% in some cases (Thompson and Willson, 1978). Most plant species ripen fruit quickly and rely on rapid removal of fruits by migrating frugivores. The relatively few plants that fruit earlier in summer when avian numbers are low and birds are concentrating on insects (Morton, 1973) tend to fruit asynchronously, thereby making only a few fruits available at any one time (Thompson and Willson, 1979). These plants also often have bicolored fruit displays, with their fruits changing in color over time from green to a conspicuous preripe color such as red and finally to a ripe color such as black, for example, wild black cherry. Others use bright colored pedicels as contrasting color to their fruits. These bicolored displays may function to increase the effective size of the fruit display thereby attracting more dispersal agents but without making so many fruits available at any one time that the resident birds are satiated. These bicolored displays may also attract fledgling birds who are still learning color cues for food resources (Willson and Thompson, 1982).

The kinds of fruiting patterns that characterize temperate plants with fleshy fruits occur also in tropical forests, but the range of alternative patterns is broader in the tropics. Studies on tropical fruiting plants over the past decade have focused on understanding how selection pressures differ among plants that rely on a diversity of visitors and those whose fruits are accessible to fewer species of larger frugivores. Snow (1971) notes that in the tropics, fruits of plants in secondary forests tend to be small and succulent, to ripen quickly, and to have many seeds, whereas fruits of trees in primary forests tend to be larger and drier, to ripen asynchronously, and to have one seed. He suggests that seed size as an adaptation to the types of habitats in which the seeds germinate is the primary selection pressure on these plants and the mutualisms with birds are built on these primary adaptations. Selection for large seed size favors in turn the evolution of a thin nutritious fruit: a thick fruit around a large seed would demand a very large frugivore, whereas a poorly nutritious fruit would not compensate for the large seed ballast and would probably be avoided by birds. Howe and Vande Kerckhove (1980) show that seed ballast can be important in the probability that a fruit is eaten. Frugivorous birds feeding on *Virola surinamensis* in Panama preferentially remove fruits in which the ratio of aril to seed is high. Therefore, selection for large seeds and dispersal by birds selects simultaneously for a small set of frugivores at the upper end of the sizes for frugivorous birds, and the nutritious nature of these fruits allows these birds to be more completely frugivorous than most small frugivores.

The important point regarding the evolution of mutual dependence is that if this scenario is correct, then selection on the host plant is not strictly for a particular dispersal agent or small group of dispersal agents that are specific to the host plant. Rather, selection is often primarily for seed size, and the few avian species that can specialize on these large-seeded fruits are a consequence of that selection. There is no indication that the large-seeded plants are any less under selection to maximize the number of visits by dispersal agents that can feed on seeds of that size than plants with small seeds and fruits—the subset of these larger frugivores is simply smaller than the subset of the smaller frugivores.

The alternative argument is McKey's (1975) suggestion that the continuum from small-seeded succulent fruits to large-seeded nutritious fruits represents primarily a continuum of selection on hosts for specificity in their dispersal agents: attraction of opportunistic dispersal agents at one end and visitors specialized for frugivory at the other end. In this view selection on hosts for opportunistic visitors is selection toward maximizing the number of visits of potential dispersal agents while incurring the cost that many of the visitors are poor dispersal agents. In contrast, selection on plants for visitors more specialized for frugivory lowers the availability of the range of dispersal agents but guarantees high quality dispersal. The components of high quality dispersal in McKey's (1975) view are the reliability of visitors, the probability that an ingested seed is deposited in a condition suitable for germination, and the ability of the dispersal agent to carry relatively large seeds.

No studies have yet been completed that compare quality of dispersal along the continuum of seed and fruit traits. Such studies are sorely needed because it is not clear that visitors specialized for frugivory necessarily provide higher quality dispersal. Unlike in pollination, constancy on a fruit resource is not important to the plant, as long as there are dispersal agents available when the fruits ripen. So the requirement of reliability of dispersal agents is much less restrictive than the requirement of constancy in pollinators. The importance of this component of dispersal quality to the plant will vary with the size of local populations of frugivores. Moreover, obligate frugivores may not necessarily deposit a greater number of seeds safely in sites appropriate for germination than opportunistic dispersal agents: the floors of the caves inhabited by oilbirds are littered with seeds (Snow, 1962c). The ability of larger frugivores to carry large seeds may be the component of quality of dispersal that intersects with the alternative hypothesis that selection is primarily on seed size and that attraction of the subset of large frugivores is a consequence of that selection rather than the actual basis of selection for large seeds.

Of course, these hypotheses are not necessarily mutually exclusive. Any theory on the evolution of interactions between birds and fruits must incorporate predictions on the ecological conditions favoring hosts that attract one type of visitor rather than another. Analyses of how different fruiting patterns maximize visits by different subsets of available frugivores (Howe and Estabrook, 1977) are an important step in this direction. The data of Howe and colleagues for tropical trees at both ends of the continuum of seed and fruit size indicate that trees with small fruits and seeds attract more opportunitistic frugivores as the size of the fruiting display increases; trees with large fruits and seeds draw from a small group of resident large frugivores and increasing the size of the fruiting display beyond a certain number does not draw in increasing numbers of frugivores (Howe and Estabrook, 1977; Howe, 1977; Howe and De Steven, 1979; Howe and Vande Kerckhove, 1979). The effects of fruit quality on the kinds of frugivores attracted, however, are less clear. Neither Frost (1980) nor Sorensen (1981) find any tendency for frugivorous birds to separate in feeding patterns among plants differing in fruit quality. Frost's study is of frugivores in an evergreen coastal dune forest in South Africa, whereas Sorensen's study is of frugivores in a British woodland. Frost interprets these results together with his results showing that the majority of plant species in the community rely on a few common frugivores for dispersal as an indication that coevolution among these species is in an intermediate stage. The alternative explanation, however, is that the simple continuum from small-seeded low quality fruits to large-seeded high quality fruits is not necessarily also a continuum from low to high host specialization in birds.

CONCLUSIONS

The general impression left by the aggregate of selection pressures affecting the evolution of mutual dependence is that extreme specificity in both visitors and hosts is likely to be common only among mutualists that live in intimate association, giving the superficial appearance of a parasite–host relationship. In most other mutualisms, one mutualist is often more dependent on its partner than the other. Whether grazerlike visitors can evolve to specialize on a single host depends upon the overall life history of the visitor and how it can get through periods of the year when the host offers few or no resources. Even if selection acts on the visitor to be specific to the host, selection on the host is often to maximize the number of visits by potential mutualists and, therefore, against restriction of visits to one visitor species.

Long-term specific coevolution between particular species of grazerlike visitors and hosts, then, must be uncommon, and the ecological conditions that allow such extreme mutual dependence in particular interactions are worth close attention. Although this chapter has focused on how the kind of interaction influences specificity in mutualists, other questions and approaches are possible from the vantage point of community structure.

1.   How is selection for mutual dependence affected by the presence of many as compared with few similar hosts (visitors) in a community? Analyses of the same interaction in many different communities are probably among the most important kinds of studies that are needed now. We know very little in a rigorous way about the flexibility in interactions between species.

2.   How does host dispersion affect selection on mutual dependence? Some species are predictably abundant and widespread in communities, whereas others are uncommon and distributed patchily. Uncommon hosts may have to invest more in a mutualism in order to make it selectively

advantageous for visitors to search out such hosts, and this could select for hosts that are more dependent on specific visitors. Alternatively, uncommoness of the host could select for hosts that are more catholic in their acceptance of visitors. How selection acts will depend upon the type of interaction. Pollination of widely dispersed plants involves many of the plants often cited as having elaborate structures for attraction of specific groups of visitors: flowers with long corollas for attraction of long-billed hummingbirds (Feinsinger and Colwell, 1978) or orchid flowers with specific scents and shapes for attraction of bees capable of flying long distances such as euglossine bees in Central America (Janzen, 1971b). This question of host dispersion and visitor specificity mimics the similar question currently being debated over host specificity in tropical parasitoids on uncommon hosts as compared with temperate parasites on common hosts (Owen and Owen, 1974; Janzen and Pond, 1975; Rathcke and Price, 1976; Hespenheide, 1979; Janzen, 1981). The difference is that in mutualisms selection for increased or decreased specificity resulting from host dispersion can act on the host as well as on the visitor.

The general point is that we have only begun to ask rigorously the kinds of questions that can indicate and explain patterns in the evolution of mutual dependence. Both detailed studies of the selection pressures affecting mutualisms and innovative ways of analyzing the evolution of mutualisms in a community context are needed to understand why mutualisms vary in the degree of mutual dependence of the interacting species.

CHAPTER

7

INTERACTION AND SPECIATION

There are at least as many ways to consider patterns in speciation as there are subdisciplines in evolutionary theory. Emphases vary from biogeographical and populational approaches on the likelihood of speciation in allopatric, parapatric, and sympatric populations (Mayr, 1970; Bush, 1975; Endler, 1977) to paleobiological approaches on phyletic gradualism as compared with punctuated equilibrium (Eldredge and Gould, 1972; Gould and Eldredge, 1977); molecular approaches range from the relative importance of chromosomal as compared with allelic changes (White, 1968, 1978; Grant, 1971; Stebbins, 1971) to the importance of regulatory as compared with structural gene changes (Brittain and Davidson, 1971; Gould, 1977; Cherry et al., 1978). No single approach is adequate to understand speciation and each approach adds its own dimension to the intricacies of when, where, how, and why speciation occurs.

This chapter explores patterns in speciation through an ecological approach based on how organisms interact and coevolve. The basic problem is this: if the different kinds of antagonistic and mutualistic interactions vary in their likelihood of coadaptation as suggested in the previous chapters, are there similar patterns in the likelihood of cospeciation?

COSPECIATION

First, to paraphrase Janzen (1980a), when is it cospeciation? As with coadaptation, the key phrase is reciprocal change. Coadaptation implies reciprocal patterns of defense or mutual attraction, and cospeciation implies reciprocal speciation resulting from an interaction between two or more species. If the interaction affects speciation in only one of the taxa, that is interesting evolutionarily but it is not cospeciation. As Janzen (1980a) argues forcefully, the powerful concept of coevolution will lose much of what it can tell us about how organisms respond evolutionarily to interactions if the term becomes synonymous with one-sided evolution in response to interaction.

The most common misuse of the concept of cospeciation is its application to congruence of phylogenetic trees of interacting species without any indication of reciprocal influences on speciation. There are assuredly many instances in which, for example, parasites speciate along with their hosts but have no reciprocal effects on host speciation, but this is not cospeciation. Nevertheless, at least one recent definition of cospeciation allows explicitly the use of the term without any reciprocal effects on speciation. When considering the concept of coevolution in general and patterns of speciation in crocodileans and their digenean parasites in particular, Brooks (1979a,b) defined cospeciation as "cladogenesis of an ancestral parasite species as a result of, or concomitant with, host cladogenesis." This is a definition of congruent phylogenies, but it should not be equated with cospeciation.

The problem of equating congruent phylogenies with cospeciation is best illustrated by an example in which cospeciation is unlikely to be occurring. Cactophilic yeasts in the genus *Pichia* live in the rotting stems of columnar cacti in southwestern United States and northwestern Mexico. The species of yeast show genetic changes between populations living on different cactus species, and some varieties seem in the final stages of speciation (Starmer et al., 1980). There is partial congruence in the phylogenies of the cacti and the yeasts, but unless these yeasts are somehow affecting significantly the living tissues of their cactus hosts, the yeasts are unlikely to be a selective force in cactus speciation. The congruence of phylogenies is a result of host selection by the yeasts (as mediated by their *Drosophila* vectors): the secondary chemistry differs between the cacti species along phylogenetic lines and the yeast species separate onto their cacti hosts accordingly. If cospeciation is used to describe this pattern of phyletic tracking of yeasts on cacti, then yet another term will be needed to describe reciprocal effects on speciation.

TEMPO IN COSPECIATION

There is no reason to expect that cospeciation must proceed at a lockstep tempo or that each speciation event in one of the species will automatically result in speciation in the other species. Rather, the more reasonable assumption is that interactions are likely to result in differential effects on speciation. In a broad sense, this is how Ehrlich and Raven (1964) envisage the process of coevolution and, specifically, cospeciation between plants and insects, using butterflies as an example: butterflies select for plants with novel chemistry that are avoided by butterflies; these plants, now free of herbivores, spread widely and radiate in species; this abundant new resource is colonized by a group of butterflies that can cope with these compounds and these butterflies in turn radiate in species; this radiation of butterflies selects again for plants with novel chemistry, and so on. In this scheme the plants influence many of the speciation events in their phytophagous insects whereas the insects are responsible for fewer, although critical, speciation events in plants. Because of this imbalance in speciation events attributable to the interaction, the impression that results from comparing phylogenies may often be one of strict phylogenetic tracking of the herbivore on plant speciation with no easily discernible effect of the herbivore on plant speciation.

If differential rates of speciation are a common occurrence among interacting species, then the study of cospeciation becomes a two-part problem: what kinds of interactions are likely to result in cospeciation, and what patterns are there in the tempo of speciation among interacting organisms? The question of patterns in speciation rates attributable to differences in how organisms interact with other species is important even outside the context of cospeciation. This question is particularly critical when considering cospeciation, however, because our assumptions on how the pattern of cospeciation should look can influence importantly our impression of the commonness of cospeciation. If we expect each speciation event to automatically generate speciation in the other species in an interaction, then we will dismiss most cases of cospeciation where the tempo of speciation differs between the species. For this reason, the following sections focus as much on how different kinds of interaction affect speciation rates as they do on the question of what kinds of interactions are likely to result in cospeciation.

ANTAGONISM AND SPECIATION

There are few studies that have actually generated plausible scenarios for how specific parasites, grazers, or predators could have induced speciation in their hosts or prey thereby leading to several rounds of cospeciation. The interactions between *Passiflora* vines and *Heliconius* butterflies seem to be the most promising among those studied thus far for demonstrating actual cospeciation. The system has all the right ingredients: (1)

Heliconius larvae are locally specific to particular *Passiflora* species (Benson et al., 1975; Benson, 1978); (2) these larvae are the major herbivores of their host plants and form the basis for an elaborate associated food web (Gilbert, 1977); (3) the plants have traits that seem to function specifically as defense mechanisms against these butterflies (Gilbert, 1975; Williams and Gilbert, 1981).

One of the most important of these defensive traits for *Passiflora* speciation may be leaf shape. *Heliconius* butterflies search for host plants visually and *Passiflora* plants with novel leaf shapes could potentially escape attack, at least to a point. Gilbert (1975) suggests that the maximum number of distinct leaf shapes that could function effectively in a single habitat is limited to the number of shapes that *Heliconius* individuals perceive as distinct. These and other *Passiflora* traits indicate that *Heliconius* butterflies are important selective agents on these plants and may be responsible in a major way for differentiation among *Passiflora* populations. Much of *Heliconius* speciation is thought to have occurred during the Pleistocene when forested areas in South America were reduced to islands surrounded by grasslands (Brown et al., 1974; Brown, 1976). The interactions between *Heliconius* and *Passiflora* species must certainly have differed from one forest island to another and differentiation of populations in both taxa is a plausible result.

These cospeciation events may represent diffuse coevolution in the sense that the interactions involve a set of *Passiflora* species and *Heliconius* species, with host switching in *Heliconius* among the *Passiflora* species. Benson et al. (1975) argue, for example, that speciation in *Heliconius* indicates a pattern of radiation in species over the increasingly derived species of *Passiflora* but with some reradiation of *Heliconius* species back onto more primitive *Passiflora* subgenera. Therefore, the basic unit of interaction in analyzing the pattern of cospeciation in *Passiflora* and *Heliconius* is the group of species within each of these genera.

Other well studied examples of potential cospeciation among parasites, grazers, and predators and their hosts or prey are few. The suggestion by Nault and DeLong (1980) that *Dalbulus* leafhoppers cospeciated with maize and its ancestors seems enticing. The spiroplasmas, mycoplasmas, and viruses that attack these plant species via transmission by the leafhoppers differ in either their occurrence over the elevational range of the plants or in their virulence to different genotypes of the plants. Nault and DeLong suggest that these differences may have been important in influencing the past distribution of maize and its ancestors and speciation in these plants. They imply that the leafhopper speciation was, in turn, a result of radiation in the plants. This suggestion of cospeciation would be worthwhile pursuing particularly because the argument involves cospeciation mediated by a third group of organisms. Some other examples of potential cospeciation are too diffuse and the taxa now too diversified to make any significant headway in studying patterns in cospeciation among particular species (hypsodent-toothed mammals and grasses, e.g.) except perhaps in some isolated situations, although broad patterns of coevolution seem evident (e.g. Stebbins, 1981).

Because few potential examples of cospeciation among antagonists have been studied in detail, there is little of a solid base from which to generate predictions on relative speciation rates among interacting parasites, grazers, or predators and their hosts or prey. The focus on rates in these interactions has come from another direction—the comparison of speciation rates between parasites and predators. Such a focus could be useful in understanding cospeciation, however, because any discernible patterns could provide a basis for evaluating how hosts or prey differ from enemies in speciation resulting from interactions. The most general hypothesis is Price's (1977, 1980) argument that rates of speciation are higher in parasites than in predators. Grazers could be included readily with predators here. The argument is based primarily on the following aspects of parasite biology: (1) parasites are often highly host specific and (2) parasite populations are often small and localized. Under these conditions, local differentiation of populations is possible either through host race formation in sympatric populations or through allopatric differentiation.

Broad comparisons among the largest families of British insects support the hypothesis. Price's (1980) comparison of the 10 largest families comprised mostly of predators, parasites of plants, or parasites of animals shows that the number of species in predatory families is, on average, eight times less than in plant parasitic families and 10 times less than in animal parasitic families. Families with grazerlike insects were not included in the analysis, but these families each have few species and are, therefore, more similar to predators in the British insect fauna than to parasites: among the 10 families of Orthoptera, the mean number of species per family is three and the range is 1–11 (based on species list in Kloet and Hincks, 1964).

Not all of the many speciation events among parasites involve speciation within closely related taxa of hosts. For example, almost every microlepidopeteran family has radiated in species to include a wide array of plant families as larval food plants (Table 7.1). Nevertheless, many other instances of speciation in parasites do follow closely host lineages (Ehrlich and Raven, 1964; Holmes, 1973; Benson et al., 1975; Brooks, 1979b; Price, 1980), suggesting that parasite speciation is often driven by

Table 7.1.
Total Number of Plant Families Recorded as Hosts for Larval Ditrysian
Microlepidoptera[a]

Taxon	No. Moth Species	No. Host Plant Families
Tineidae	300	6
Lyonetiidae	215	50
Gracillariidae	755	79
Elachistidae	104	7
Coleopridae		
Coleophorinae	250	34
Cosmopterygidae	222	39
Gelechiidae	757	82
Momphidae	36	8
Oecophoridae		
Ethmiinae	58	10
Stenomatinae	65	16
Xyloryctinae	114	21
Depressariinae[b]	165	17
Other	453	33
Blastobasidae	56	12
Scythrididae	61	20
Carposinidae	37	19
Tortricidae		
Olethreutinae	841	86
Torticinae	540	81
Cochylidae	104	27
Plutellidae and		
Yponomeutidae	191	55
Argyresthiidae	62	13
Choreutidae	50	17
Sesiidae	169	40

Source: from compilations in Powell (1980).
[a] Data are divided into moth subfamilies for a few of the largest moth families and for families with adequate data for only one of several subfamilies. Moth families with few host records are excluded. Number of moth species refers to the number of species with recorded hosts.
[b] Only Depressariini.

host speciation. In fact, the circumstances that allow radical host shifts may have a predictable pattern that results from mating away from the host plants. Gilbert (1979) shows that radical shifts of butterflies onto chemically distant plant families are more common in butterfly families that mate near their host plants than in those families that do not mate near their host plants. That is, the more closely associated a butterfly species is to its host plant through all of its life history stages, the more likely the chance that a radical shift in host plants will result in speciation.

If the parasitic way of life is especially prone to speciation, the complementary question from a coevolutionary perspective is whether hosts interacting mostly with parasites are more likely to speciate than hosts or prey interacting with grazers or predators. If local differentiation in parasites is generally more likely than in grazers or predators, and if such local differentiation can impose selection for similar differentiation in hosts, then parasitism may generate higher rates of speciation in hosts than either grazers or predators. At this point it is not possible to answer these "if" statements; but they are worth posing as hypotheses to be tested because we know so little about how the modes of interaction between species can potentially affect speciation.

All of the above discussion has centered on cospeciation in antagonistic trophic interactions. As in coadaptation, cospeciation of competitors is a special situation. It is conceivable that different populations of the same competitor species could coadapt in different ways so that differentiation and reproductive isolation occurs. But this simply returns us to the related problem that most adaptation resulting from competition is likely to be one-sided and that there is seldom a selective force to keep the interactions together (Chapter 3). If cospeciation does occur among competitors, it seems most likely that it would happen in the context of competition for mutualists such as pollinators of plants, where it becomes somewhat of a semantic question as to whether the basis of cospeciation is the competitive or the mutualistic interaction. There have been no detailed studies of potential cospeciation among competitors.

MUTUALISM AND SPECIATION

Mutualisms would seem a priori to result in more spectacular instances of long-term cospeciation than antagonistic interactions because the positive effects of these interactions on the fitness of both (all) participants would hold the interactions together through speciation events. Most mutualisms, however, lack the host and visitor specificity necessary to make this attractive idea plausible. The most likely examples would be among mutualisms where the visitor lives in intimate association with the host, but some of the best known examples do not seem to indicate repeated instances of cospeciation within lineages. Instead there is evidence for a process involving collection of new species of hosts or visitors from a series of lineages, and this may be a common speciation process among mutualisms. For instance, the 12 species of swollen-thorn acacias in Central America are associated with at least 13 species of *Pseudomyrmex* ants (Janzen, 1974b). The ants are all derived from one section of the genus *Pseudomyrmex* except for *P. nigropilosa*, which lives in acacias but is not mutualistic (Janzen, 1975a). The swollen-thorn acacias, however, do not represent a single phyletic lineage but rather may be derived from as many as five separate acacia lineages (Janzen, 1966). The ants are not specific to single acacia species, but instead are specific to the life form: a single ant colony may occupy two adjacent acacia individuals of different species (Janzen, 1974b). Since the initial mutualism developed, the ants have apparently collected unrelated acacias with similar traits.

Similarly, the ant-fed plants of Malaysia and New Guinea represent a series of unrelated plant species from several plant families that have been collected into the interaction by at most a few species of *Iridomyrmex* ants. In fact, many of these plants are associated with what may be only one species of *Iridomyrmex* (Janzen, 1974c; Huxley, 1978). These interactions may not have generated any speciation in the ants and the number of plant species in this mutualism involves both speciation within lineages and a collection of a group of unrelated lineages.

These examples indicate (1) that mutualistic interactions do not always build in species numbers through speciation within lineages, and (2) that the tempo of any cospeciation that may occur through these interactions may vary greatly among the participant species. These two processes also seem apparent in interactions among birds and fruits. Snow's (1971, 1981) arguments on how bird–fruit mutualisms affect speciation imply differential effects of these interactions on bird and plant speciation, although he does not develop his arguments exactly in this context. Snow's (1971) original hypothesis was developed as a comparison of speciation rates in tropical frugivorous bird taxa as compared with related insectivorous taxa. He argues that utilization of fruits by birds favors plants that produce abundant supplies of easily accessible and exploitable fruits that do not require special hunting techniques. In contrast, utilization of insects has selected for a heterogenous, sparse, and cryptic distribution of those prey. Effective utilization of insects as prey, therefore, has selected for birds with many different hunting techniques. Snow suggests that one consequence of this difference between fruit and insects as food is that families of insectivorous birds tend to have more species than families of frugivorous birds. Among four neotropical suboscine passerine families, the mainly frugivorous Cotingidae (cotingas) have 90 species and the Pipridae (manakins) 59 species, whereas the mainly insectivorous Formicariidae (ant-birds) have 222 species and the Tyrannidae (tyrant flycatchers) 365 species (Snow, 1971).

The effect of bird–fruit mutualisms on plant speciation, however, may differ from the effect on avian speciation. At the most general level, Regal (1977) suggests that the explosive rise to dominance of the angiosperms resulted from both insect pollination and seed dispersal by birds. Together these mutualisms allowed angiosperms to reproduce successfully even though they were not near to conspecifics. A seed dispersed to a disturbed area away from conspecifics could still reproduce because insects could carry pollen to and from the plant to conspecifics growing outside the range in which wind-dispersed pollen is usually borne. But what about cospeciation among particular birds and plants? Adopting a more specific tack, Snow (1981) argues that occasional long-distance dispersal of bird-dispersed seeds may establish plant populations away from the parental species and that after a period of differentiation, individuals from the daughter population might reenter the range of the parental species as a new species. He argues further that since the fruits of the parental species and the new species would likely still be similar enough so that they would draw on the same frugivore assemblage, competition may favor the separation of fruiting seasons between these species.

Even if the colonizer population did not reenter the range of the parental species as suggested in Snow's scenario, however, the mutualism is what provides the raw material for potential speciation in the plant. Dispersal of seeds away from the parental population can generate allopatric populations that over time become good biological species.

Combining the arguments on patterns of speciation in frugivorous birds and bird-dispersed plants, the expected general result may be that bird–plant mutualisms decrease speciation rates in birds and increase speciation rates in plants relative to closely related species not involved in bird–fruit mutualisms. The limiting partner in cospeciation among birds and plants, therefore, is probably the birds. It is clear how the dispersal activities of the birds can influence plant speciation. How often cospeciation actually occurs, however, depends upon how often speciation in frugivorous birds is a result of their interactions with different populations or species of fruiting plants.

It may be that the tempo of cospeciation among birds and plants is the reverse of that envisioned by Ehrlich and Raven (1964) for insects and plants. The birds may be responsible for many instances of speciation in the plants, whereas the plants may be responsible for fewer, although critical, instances of speciation in the birds generating deviant groups within avian families or new families whose species have life histories closely tied to the exploitation of fruits: the monospecific oilbird family Steatornithidae and some speciation events in the Pipridae, Cotingidae, Trogonidae, Columbidae, and Paradisaeidae, for example. The nonspecific nature of these mutualisms must certainly allow unrelated lineages to be collected into these mutualisms over evolutionary time.

IS POLLINATION A SPECIAL CASE?

The search for generalizations in patterns of cospeciation among mutualists is certainly still in an early stage of analysis, and in the process the better-known mutualisms are likely to serve as models. Since interactions between plants and pollinators are probably the best studied mutualisms, these interactions are likely to become the models for studies of other mutualisms. This could be dangerous, however, because these mutualisms may be poor models. Wheelwright and Orians (1982) caution that the terminology and conclusions drawn from patterns of coadapatation in pollinator–plant mutualisms are being applied to bird–fruit interactions without testing whether the transfer of conclusions is warranted. Their caution is applicable as well to comparisons of pollination with other mutualisms in the likelihood of cospeciation. In no other kind of mutualism does either the visitor or the host have such direct control over reproductive isolation as does a flower visitor pollinating a plant. The pollinator controls with whom a plant does and does not mate, and such direct control over mate choice in a mutualist may be unique among mutualisms.

Pollination by animals may result in cospeciation more often than other kinds of mutualism. Also unlike many other mutualisms, flower visitation by animals often favors local specificity of the visitor to the plant from the viewpoints of both the visitor and the plant. Constancy of the visitor to one plant species increases its efficiency at harvesting pollen or nectar (Heinrich, 1979), and such constancy increases the likelihood that the pollinator deposits conspecific pollen on the stigma of the next flower it visits. The pattern of movement of the flower visitor can determine to an important extent the effective size of the plant population (Levin and Kerster, 1974; Levin, 1979; Schmitt, 1980), hence the likelihood of differentiation among plant populations. Simple allelic changes such as differences in floral color can influence the likelihood of being visited by the local assemblage of pollinators (Waser and Price, 1981), and it is conceivable that such differences among plant populations could result in plant speciation. To the degree that pollinators are specific to the traits of their local plant populations, changes in floral traits can lead to pollinator speciation as well. Cospeciation in these circumstances may be essentially concomitant.

For example, speciation in *Nemophila* spp. (Hydrophyllaceae) and the oligolectic *Andrena* bees that visit these plants seems to represent cospeciation mediated by the pattern of floral visits in the bees and is based on a level of visitor and host specificity uncommon among other mutualisms. The *Nemophila* species in different parts of California vary in their reflectance patterns and phenology and the andrenid pollinators are for the most part specific to their particular *Nemophila* hosts (Cruden 1972a,b). The differences in reflectivity in the flowers are likely to have occurred when the *Nemophila* populations were separated during the Miocene; current distribution of *Nemophila* plants with different reflectivity follows well the known past geological history of the region. The bees in the different regions are the most likely selective agents for these differences in the flowers and the host-specific tendencies of these bees are likely to have maintained the differences after the populations came back into contact. The process of speciation in both the plants and the bees has most likely been the result of their interaction. Similarly, White (1978) argues that speciation in figs and wasps must be concomitant. It may be impossible to separate the independent variable in speciation in these interactions between flowers and pollinators where the potential for speciation in the plant is so directly under the control of pollinator. These interactions probably serve as counterpoints to the process of speciation in other mutualisms rather than as appropriate models for those mutualisms.

CONCLUSIONS

This chapter has been intended as an introduction to a way of searching for patterns in speciation that has been the province of relatively few researchers interested in modes of interaction between species and in coevolution. The major points have been as follows:

1. Rates of speciation may depend upon how species interact with other species.

2. Tempo of speciation may vary between species in the same interaction.

3. Increase in the number of species in a particular mutualism may often result from the collection of new unrelated hosts or visitors rather than through speciation within lineages.

4. Flower–flower visitor mutualisms may be poor models for analyzing cospeciation in other mutualisms.

Together these statements provide an approach to the study of speciation that may complement the more traditional biogeographic, paleobiological, and molecular approaches.

The extent to which cospeciation—in the sense parallel to coadapation—varies among modes of interaction between species is a subset of the question of how interactions affect the tempo of speciation. If interactions have differential effects on speciation in, say, mutualistic visitors and hosts or parasites and hosts, then any actual cospeciation will be unbalanced between the interacting taxa. How coadaptation and cospeciation are related evolutionarily is part of this problem. The emphasis in this chapter has been on cospeciation as a result of differential coadaptation among different populations of the interacting species. Other approaches are possible but, without coadapation as part of the process, the results will probably not fit comfortably into the strict definition of cospeciation.

Perhaps the central question in analyzing the influence of interactions on speciation and of the likelihood of cospeciation is simply a question of what causes reproductive isolation of populations: how often is speciation a result of selection acting differently between populations because of differences in their interactions with other species. The point of this chapter has been to suggest that in answering the question, a careful analysis of the mode of interaction and the relative tempo of speciation among interacting species may offer some useful direction.

CHAPTER

8

THE INTERACTION STRUCTURE OF COMMUNITIES

Most of what is interesting about biological communities cannot be pinned, stuffed, pressed onto herbarium sheets, or preserved in alcohol. Knowing the species structure of an assemblage of organisms tells us in and of itself little more than a telephone book tells us about a city. Nor can what is interesting about biological communities be dissected, weighed, separated on starch gels, or centrifuged into supernatant and precipitate fractions. Knowing the internal workings of organisms in isolation from other organisms with which they interact tells us the "how" of life without the "why." What makes biological communities more than lists of taxa complete with details of how they tick are the interactions among species.

This chapter is entitled the interaction structure of communities because I focus here on four aspects of interaction that are meaningful only in a community context. I develop each of them as problems in need of field experiments rather than as statements of known ecological patterns. Each of the questions has appeared in one form or another in previous chapters. These questions, however, are collected here in a broader context to emphasize how analyses of interaction and coevolution together represent a coherent way of studying community organization, complementary to other approaches such as those based on species diversity or energy/nutrient flow.

The questions are these:

1. What are the units of interaction and coevolution within communities?
2. How do interactions grow in associated species over evolutionary time?
3. What are the limits of the ripple effects of interactions throughout communities?
4. Are there patterns to the demography and geography of interactions?

THE EVOLUTIONARY UNITS OF INTERACTION

In Chapter 5 I argued that the evolution of mutualisms is often favored because of interactions between at least one of the mutualistic species and a third antagonistic species. The basic evolutionary unit in many of these interactions involves at least three species. If we are to understand how species coevolve, the kinds of interaction that favor coadaptation, and the kinds of interaction that produce cospeciation, we need to know the unit structure of interaction within communities for different kinds of interactions. Specifically, we need detailed studies designed to determine the unit group of species within which selection acts significantly on all participants (e.g. Gilbert, 1977; 1979). Until we have a broad sampling of such studies we cannot know how much our views on coevolution are biased by the preponderance of studies of species pairs analyzed out of context with other interactions in a community.

Approaches to understanding the interaction structure of communities have come from several directions of ecological emphasis over the past decade or so, and each of these approaches tells us something about the unit of interaction. Root's (1967, 1973) concepts of guild and component community provide important heuristic tools for studying how interactions are subdivided within communities. Root defines guild as a group of species that exploits the same resource in a similar way (Root, 1967), and he defines component community as an assemblage of species associated with some microenvironment or resource (Root, 1973). In Root's studies, guilds have included foliage-gleaning insectivorous birds and sap-feeding, and pit-feeding insects on collards, and component communities have included the arthropod assemblage on collards. Although neither of these concepts directly suggests an approach to interaction based on the unit

group of species within which selection acts significantly on all participants, analysis of both guild and component community structure can
suggest the subset of species that are most likely to be interacting within
communities.

Concepts similar to component community but implying a coevolved
group of interacting species include Gilbert's (1977, 1980) term coevolved
food web and Paine's (1980) term module. These are the kinds of
component communities that form the strong interactive units within
communities. In the studies of Gilbert and colleagues, the coevolved food
web based on *Passiflora* vines includes the vines, *Heliconius* butterflies,
and a beetle that feed on these plants, parasitic wasps that attack the eggs
and pupae of *Heliconius*, *Anguria* flowers that provide pollen for the
long-lived *Heliconius* adults as they search for *Passiflora* meristems on
which to oviposit, and the complex of other insects associated with
Anguria (Gilbert, 1977).

These units of interactions are often characterized further by what
Paine (1980) calls strong interactions among the species. That is, although
species may interact with many other species in a community, the
essential unit for which the ecologist searches is that unit within which
removal of the species would have significant effects on the other species.
In Paine's (1966, 1969, 1974, 1977, 1980) experiments these effects are
population changes in the other species and these elegant experiments
have been critical in our understanding of the ecological units of interaction.

Often the species that have strong effects on the population levels of
other species will be the same as the species critical in the evolutionary
unit of interaction, but this will not always be the case. For example,
some mutualisms may have no effect on the population levels of interacting species. A mutualism is favored by selection because it allows those
individuals possessing traits that foster the interaction to increase their
genetic contribution to future generations relative to other individuals in
the population. The mutualism may have little or no effect on the overall
population levels of the species. Experiments designed to determine the
unit of interaction within which selection acts significantly on all the
species cannot use changes in population levels resulting from a manipulation of one of the species as the sole criterion of the limits of the important
species. Natural selection working through interactions among species
acts through the relative effects of interactions on survival and reproduction of individuals engaged in a particular interaction and not on the effect
of an interaction on overall population levels.

Deciphering the evolutionary units of interaction tells us more than
how the interaction structure of communities is assembled. These

analyses can also indicate the constraints under which coevolution proceeds between particular pairs of species. In the same sense that particular adaptations in organisms are constrained by architectural features of body plans (Gould, 1980), coevolution among particular pairs of species is constrained by the conflicting selection pressures imposed by other organisms with which these species interact. "Perfect" adapation is limited by these constraints and delimiting the broader set of species that encompass the evolutionary unit of interaction is the first step in specifying the constraints.

The tiny moth *Greya subalba* (Incurvariidae) and its *Lomatium* (Umbelliferae) host plants, for example, seem amazingly inept at attack and defense. The interactions between these species are comprehensible only when placed in context of the other species that are associated closely with these interactions and form the evolutionary unit of interaction. The moth oviposits in the immature *Lomatium* seeds in the steppe of the northwestern United States and the larvae feed within the seeds. The seeds are borne on umbels similar in appearance to wild parsnip *(Pastinaca sativa)* and wild carrot *(Daucus carota)*. When an adult female *Greya* finds a host plant, she lays eggs in only a few of the seeds on each umbellet even though most of the remaining seeds usually are unattacked and a female still has more eggs ready for oviposition. Therefore, *Greya* females expend energy that could be used potentially for egg production instead of searching for new host plants.

This behavior of placing only a few eggs among a large group of available seeds seems to result from selection to minimize losses to parasitoids that search for larvae one seed at a time (Thompson, unpublished data). When a parasitoid or small predator lands on an umbel, it cannot determine readily which and how many seeds have larvae and, therefore, must probe seeds in search of larval hosts. Finding a larva in one seed provides little or no information on the likelihood of finding another larva in a nearby or any other seed.

Some *Lomatium* populations are also attacked by a seed parasitic beetle. *Lomatium* individuals would suffer even lower rates of attack on seeds by seed parasites if they did not elongate their flower pedicels between the time of flowering and the time of seed dispersal. Fewer seeds are attacked in umbels that are manipulated experimentally so that seeds remain in close contact until the time of dispersal. Such a defense, however, although readily available in *Lomatium,* would lower the rate of seed dispersal as occurs when the plants are manipulated to simulate such a defense (Thompson, unpublished data). Since seeds are usually dispersed from early to mid-summer, those remaining on the plants remain easily locatable food sources for hemipteran insects that suck nutrients from the seeds, although this aspect of the interaction has not yet been quantified.

Hence a *Greya* female cannot exploit fully a *Lomatium* host plant she discovers and *Lomatium* individuals cannot utilize a readily available potential defense because of conflicting selection pressures resulting from their interaction with other species. Any potential coevolution between *Greya* and *Lomatium* seems constrained by these other interactions. Such constraints on coevolution among pairs of interacting species are probably common and argue strongly for studies designed to delimit groups of species that interact as an evolutionary unit within which selection on any one pair influences the interactions among the other species.

THE GROWTH OF INTERACTIONS

The evolution of an interaction between two species can become a focal point around which other species evolve and become part of the interaction. Some of the growth of interactions is by cospeciation but much of the growth is through the collection of unrelated species that take advantage of resources made available by the interaction. This natural augmentation of interactions over evolutionary time is best developed among mutualisms rather than among antagonistic interactions because the evolution of mutualistic interactions often involves the development of some easily exploitable resource. In contrast, antagonistic interactions such as those between herbivores and plants seldom result in the development of new readily available resources and are, therefore, unlikely to collect many additional species simply because the interaction occurs. Plant galls are probably the major exception to this suggestion, since galls are new resources created through the interaction of plants and gall-forming species.

This difference between antagonism and mutualism needs further explanation. Antagonistic interactions, of course, can grow within communities over evolutionary time, particularly as new resources are added. The fast rate at which newly introduced plants are sometimes colonized by phytophagous insects is indicative of the speed with which antagonistic interactions can begin to develop once a new resource is added to a community (Strong, 1974; 1979). However, the development of these new interactions results from other interactions only in the sense that the phytophagous insects are already in the community because they are feeding on other plants and because the introduced plant carries with it the chemical, morphological, and behavioral traits that were selected by its past interactions with herbivores; these traits now limit which of the insects can effectively colonize the plant in its new habitat. In at least some mutualisms, however, a new resource is developed through the interaction and becomes available to other species because mutualisms often do not favor high levels of host and visitor specificity. The major new resources created through mutualisms in terrestrial communities include extrafloral nectar, floral nectar, fleshy fruits, and other dispersal structures on plants.

The evolutionary result is that mutualisms can become links with a wide variety of organisms that are otherwise part of very different component communities. Parasitoids associated with a broad array of insects on many different host plants visit the flowers of umbellifers, since many of these plants have nectar easily available to short-tongued insects (Leius, 1960; Bell, 1971; pers. obs.). These visits to flowers are known to increase adult longevity and/or fecundity in some parasitoids (Leius, 1961a,b, 1963). Although the parasitoids are probably not generally the pollinators of the flowers they visit, the development of the mutualism between flowers and other insects has allowed exploitation of the mutualism by the parasitoids. In this sense, the plant family Umbelliferae may be a critical link between component communities in temperate steppe and grassland communities where this family is most common, despite the fact that umbellifers are seldom among the dominant species in terms of biomass or cover within communities. Leius (1967) notes that the presence of Umbelliferae in some habitats increases the population levels of parasitoids within local areas, as indicated by parasitism levels on host insects. Gilbert (1980) refers to such species as keystone mutualists and uses an umbellifer, wild carrot *(Daucus),* among his examples.

The development of mutualisms can also favor the evolution of species that specialize in exploiting the mutualism for other purposes. Flowering plants can become focal points for predators in search of prey. Crab spiders, phymatid bugs, and mantids all include taxa specialized for exploiting plant−pollinator mutualisms in order to obtain prey. Although some spiders use webs or trap doors in order to catch any prey that happen to fly or crawl nearby, crab spiders use flowers as a means of increasing the likelihood that prey will come within reach. These taxa of crab spiders, phymatid bugs, and flower mantids exist because of these mutualisms. Similarly, fruiting trees can become focal points for predators such as falcons in search of prey (Howe, 1979). It would not be surprising if some populations or species of raptors were specialized to search for prey by keying in on fruiting trees throughout their home range.

The growth of interactions around mutualisms is not all a result of exploitation of the mutualisms. Much of the growth can occur through the collection of new mutualists who use the mutualism during part of their lives or become totally dependent on the mutualism. Species of birds from a wide diversity of families utilize fruits as an important component of their diet in both temperate and tropical latitudes (Thompson and Willson, 1979; Snow, 1980, 1981), and many of these birds are probably effective to some extent at dispersal of seeds. To become part of these mutualisms in at least a minor way probably involves little modification in the behavior and morphologies of birds. As a consequence, these mutualisms could easily continue to grow over evolutionary time. In addition, any plant that evolved fruitlike structures now that there are many partly frugivorous species already within communities can probably be absorbed into these mutualisms.

In this way the evolution of mutualisms can be an accelerating process in communities. Species with newly evolved traits or species that colonize communities that already have similar mutualisms can be rapidly incorporated into the local mutualist assemblage. Such rapid incorporation is most likely among mutualisms where specificity of hosts and visitors is low, although the likelihood of development of the mutualism will depend upon the local assemblage. *Plantago lanceolata* introduced into Queensland is visited frequently and is probably bee-pollinated in Queensland but is wind-pollinated in Britain (Clifford, 1962).

In some mutualisms specialization of a lifestyle to a mutualistic interaction is not possible until a broad array of species has been incorporated into the mutualism within a community. Oilbirds and other obligate frugivores are possible as evolutionary products because many species within the communities in which they evolved had fleshy fruits that allowed these birds to feed on fruits year round. Similarly, the lifestyles of social bees require a seasonal progression of flowers from which they can collect pollen and nectar. Therefore, obligate specialization to a particular kind of mutualism at least in visitors sometimes requires that the number of host taxa collected into the mutualism has passed some threshold level.

The point of this discussion is that mutualisms trigger development of related interactions that involve either exploiting these interactions or diffusing the mutualisms through more species in the community. These observations emphasize certain worthwhile directions in the study of the interaction structure of communities: (1) the search for patterns in how interactions produce new resources that become the focal points for collecting new species into the interactions, and (2) the analysis of how mutualisms can form major links among component communities.

THE RIPPLE EFFECTS OF INTERACTIONS

We know very little about the extent to which interactions ripple in their effect through the remaining species in communities. There are species that depend upon particular interactions even though they are not involved directly in those interactions. As we continue our wholesale dismantling of communities we will see increasingly these indirect effects of interactions and we should be in continual search for them. These ripple effects of interactions can tell us much about how coevolution affects the overall interaction structure of communities.

The highly coevolved interactions between yuccas and yucca moths and between figs and agaonid wasps, for example, have ripple effects on related moths and wasps. The moth genus *Prodoxus* depends on pollination of *Yucca* by *Tegeticula* moths. *Prodoxus* larvae develop in the flower stalks or fruits of *Yucca,* and the flower stalks of at least some *Yucca* species wither rapidly if the flowers are not pollinated (Riley, 1892; Davis, 1967). Similarly, *Ficus sycomorus* has two agaonid wasp species associated with it, of which one pollinates the flowers while ovipositing and the other oviposits without pollinating the flowers. The larvae of the nonpollinator wasp species are dependent upon the pollinator for development of the fruit and, therefore, larval resources (Wiebes, 1979).

Among organisms that feed antagonistically on the same plant or animal species, a bout of coevolution between a victim species and a particular parasite, grazer, or predator can potentially influence how some or all the other species interact with the victim. Because of this link among consumer species with a particular victim, Janzen (1973) argues that all insects that feed on the same plant are ultimately in competition over evolutionary time, but they interact among themselves only indirectly through their effects on plants in contemporary and evolutionary time. No one has yet observed how an evolutionary response of a plant to a parasite, grazer, or predator affects how and if the other species can continue to feed on the plant.

Not all the indirect interactions among species feeding on the same plant species, however, are antagonistic. Needle-clawed bushbabies *(Euoticus elegantulus)* and a few other mammalian species are specialized to feed mostly on gums exuded from the damaged trunks of trees. Most of the damage seems to be caused by homopteran insects that pierce the plants with their stylets. Bushbabies forage each night along the same route, spending most of the time collecting gum plus a few insects and fruits. The gums comprise, on average 77% of the weight of foods found in the stomach and caecum (Charles-Dominique, 1977). Therefore, the bushbabies are linked indirectly with the Homoptera and other animals that damage plants with acceptable gums in a manner that causes exudation of the gums.

Still other species can be linked indirectly through a common mutualist. In the forest of the northwestern United States, both coniferous trees and some small mammals depend upon hypogeous fungi (i.e. fungi that produce their sporocarps below ground). These fungi and the conifers form mycorrhizal relationships upon which the trees depend for establishment. Some small mammals include these fungi as a major component of their diet and the fungi have traits that appear to have evolved to attract small mammals which then disperse the fungal spores in their feces (Maser et al., 1978). All these species seem to be interlinked, although some of the trees and the small mammals interact only indirectly.

The classical superorganism concept of communities, which derived from older concepts on the balance of nature (Egerton, 1973), is the extreme view of ripple effects of interactions within communities. Such views are untenable, since they suggest a hierarchy of selection and a fine-tuning of interactions beyond anything that most evolutionary ecologists consider plausible. But the more restricted view, that interactions have some indirect effects on interacting species and on a subset of the other species in a community, is reasonable and worth pursuing. Some aspects of these effects have received mathematical treatments (Levine, 1976; Slatkin and Wilson, 1979; Vandermeer, 1980). D. S. Wilson's (1980) models and discussion of selection in a community context are pointed toward evaluating some kinds of indirect effects. His specific arguments on "weak altruism" and superorganisms are controversial, but his emphasis on analyzing selection in a community context and on the potential indirect effects of interactions is important.

THE PATCH DYNAMICS OF INTERACTIONS

Coevolution is likely only where interactions among particular species are common enough and intensive enough to exert selection on those species. Therefore, the study of where interactions occur both within and among communities—that is, the demography and geography of interactions—is critical to understanding where different kinds of interactions between species are likely to result in coevolution. Furthermore, if we understand how the patterns of disturbance and change in species composition influence the interaction structure of communities over time, and how interactions vary in likelihood along environmental gradients, then we will have essentially a view of the patch dynamics of interactions in communities.

"Patch dynamics of interactions" implies the study of (1) where different kinds of interactions occur in nature, (2) the causes, sizes, and frequency of occurrence of patches of intense interactions of different kinds, and (3) how interactions change among the species at a particular place in the landscape over time. For example, we have no idea how the interactions among species change in a forest light gap from the time that the gap is created by a treefall until one or more trees eventually matures to fill the gap. Knowing how new sites for colonization are created over time, the size and frequency of those colonization sites, and the patterns of change in population and community structure over time tells us much about the patch dynamics of populations (Thompson, 1978) and communities (Pickett and Thompson, 1978; Pickett, 1980; Gilbert, 1980). The patch dynamics of interactions is an allied but additional approach to patch dynamics.

A small but increasing number of studies is focused on spatial patterns in interactions within and among communities (Table 8.1). For example, the faster removal rates of fleshy fruits from plants in light gaps as compared with plants under a closed forest canopy in the midwestern United States suggests that these sites are major focal points for interactions between temperate frugivores and plants (Thompson and Willson, 1978). Many more similar studies are needed for a great variety of interactions in order to understand how the intensity of these and other interactions varies within communities. In addition, studies of similar interactions are needed across a range of environmental gradients, disturbance regimes, and assemblages of different species, in order to separate site-specific results from general results regarding particular types of interaction. Using the bird–fruit example again, although the results on removal rates of fruits in Illinois are robust for different trials using different plant species and for different seasons, these results may not hold in all types of forests. The Illinois forests have light gaps with dense growths of vines, and the structure of these gaps differs from other kinds of forests, such as those dominated by conifers. Furthermore, the layering of plant species in these forests differs from many tropical forests, and tropical forests have fruit types that are absent from temperate forests and which are exploited by more obligately frugivorous birds.

Therefore, the Illinois experiments need to be retested in other kinds of forests and for a broad range of fruits and frugivores to see how these interactions vary in their probability of occurrence throughout communities. The results of Schemske and Brokaw (1981) on the distribution

Table 8.1.

Examples of Spatial Variation in the Probability of Interaction Between Plants and Animals in Natural Communities

Interaction or Taxa	Component Analyzed	Spatial Pattern	Reference
Extrafloral nectaries and ants	Visitation rate; protection of plants from herbivores	Forest edges and clearings > forest interior	Bentley (1976)
Temperate fleshy fruits and avian frugivores	Fruit consumption rate	Light gaps and forest edge > forest interior	Thompson and Willson (1978)
Sanguinaria canadensis L.[a] and ants	Frequency of seed dispersal	40–100 yr. old forest > 25 yr. old forest > 25–100 yr. old forest with flooding and trampling	Pudlo et al. (1980)
Cassia biloba L.[b] and			
a. Trigona fuscipennis Friese[c]	Pattern of visitation to flowers	Plants in dense patches > isolated plants	Johnson and Hubbell (1975)
b. Trigona fulviventris Guerin[c]	Pattern of visitation to flowers	Isolated plants > plants in dense patches	Johnson and Hubbell (1975)
Amyema[d] spp. and Ogyris amaryllis Hewitson[e]	Probability of attack	Sites with ants > sites without ants	Atsatt (1981a)
Senecio jacobaea L.[f] and Tyria jacobaea L.[g]	Probability of attack	Sites with Formica ants > sites without Formica ants; sunny sites > shaded sites	van der Meijden (1979)
Aristolochia[h] spp. and			
a. Battus philenor (L.)[i]	Pattern of oviposition	Sunny sites > shaded sites	Rausher (1979)
b. Battus polydamus (L.)[i]	Pattern of oviposition	Sunny sites > shaded sites	Rausher (1979)

Species	Response measured	Pattern	Reference
c. *Parides montezuma* (Westwood)[i]	Pattern of oviposition	Shaded sites > sunny sites	Rausher (1979)
Guazuma ulmifolia Lam.[b] and *Amblycerus cistelinus* (Gyllenhal)[j]	Percentage of seeds attacked	Wet sites > dry sites	Janzen (1975b)
Andira inermis (W. Wright)DC[b] and *Cleogonus* weevils[k]	Percentage of seeds attacked	Seeds below parent trees > seeds below bat feeding roosts > seeds dropped between parent tree and roost	Janzen et al. (1976)
Astragalus canadensis L.[b] and curculionid weevils	Weevils per plant stalk and flowering raceme	Low plant density sites > high density sites	Platt et al. (1974)
Asclepias syrica L.[l] and *Oncopeltus fasciatus* (Dallas)[m]	Insects per pod	High plant density sites > low density sites	Ralph (1977)
Pastinaca sativa L.[n] and *Depressaria pastinacella* (Duponchel)[o]	Probability of attack	Isolated plants > plants in dense patches; plants on patch edge > plants in patch interior	Thompson and Price (1977); Thompson (1978)
Themeda triandra Forsk[p] and *Connochaetus taurinus* Thomas[q] or *Syncerus caffer* Sparrman[q, r]	Percentage of plant eaten	Sites with low relative abundance of associated unpalatable species > sites with high relative abundance of associated unpalatable species	McNaughton (1978)

[a]Papaveraceae. [b]Leguminosae. [c]Apidae. [d]Loranthaceae. [e]Lycaenidae. [f]Compositae. [g]Arctiidae. [h]Aristolochiaceae. [i]Papilionidae. [j]Bruchidae. [k]Curculionidae. [l]Asclepiadaceae. [m]Lygaeidae. [n]Umbelliferae. [o]Oecophoridae. [p]Poaceae. [q]Bovidae. [r]Relative abundance of associated unpalatable plants had no effect on two other mammalian herbivores, *Gazella thomsonii* Gunther (Bovidae) and *Equus burchelli* Gray (Equidae).

135

of understory birds in Panamanian forests, for instance, indicate that only a few species are found more commonly in these light gaps than in the forest interior. Perhaps experiments on removal rates of fruits in Panamanian lowland forests will not show any differences between plants in light gaps as compared with plants under closed canopy.

Changes in the composition of avian guilds along environmental gradients suggest indirectly that the interaction structure differs greatly among communities. Along an elevational gradient from 400 to 3600 m in the Cordillera Vilcabamba, Peru, the number of insectivores decreases 5.1-fold, frugivores and granivores 2.3-fold, and nectarivores 1.2-fold (Terborgh, 1977). At low elevations the number of insectivorous species is vastly greater than the number of frugivorous, granivorous, and nectarivorous species, but at 3600 m all groups—with several guilds lumped into each group—have almost equal numbers of species.

The species composition of avian guilds can also change greatly between years at the same site indicating that the components of the interaction structure of communities involving avian species are probably variable in time, although no studies have yet analyzed how variation in guild structure affects interactions within communities. Karr's (1980) continuing studies of variation in avian guild structure at a site near Gamboa, Panama represent the longest series of quantitative records for a single tropical site. Over the past decade these studies have shown that avian guilds vary in species composition within guilds, and also that some guilds fluctuate in overall abundance more than other guilds.

Not only are there still few quantitative replicated studies of where certain kinds of interactions are most likely to occur within and among communities, but no studies have yet quantified how the interaction structure of a community or even a small component community changes over time. The closest attempts have been analyses of the potential change in trophic structure within communities as species composition changes (e.g. Heatwole and Levins, 1972; Becker, 1975; Simberloff, 1976), analyses of change in trophic or guild structure within component communities, over a season (e.g. Root, 1973; Lawton, 1978), or analyses of how the probability of interaction between two species changes at the same site over several years as related to the demography of the host (Thompson and Price, 1977; Thompson, 1978).

The following kinds of analyses are needed particularly:

1.   Study of changes in the interaction structure in and around areas of disturbance over time can complement analyses of changes in population levels of particular species and of changes in species composition. Disturbances are features of all landscapes and certainly affect the

interaction structure of communities. Comparative analyses of changes in interaction structure of local assemblages of species resulting from disturbances of different sizes, frequencies, and causes will be especially valuable, since these variables are known to affect differentially species composition (Dayton, 1971; Levin and Paine, 1974; Grubb, 1977; Connell, 1978; Hartshorn, 1980; Pickett and Thompson, 1978; White, 1979; Denslow, 1980a,b; Pickett, 1980).

2. Study of the differences in how the interaction structure of communities changes in response to disturbance along environmental gradients and at replicates of the same kinds of sites can help separate site-specific results from more general results.

3. Studies of how interaction structure changes within communities over time at sites subjected to minimal disturbance provide a critical control. How much season-to-season and year-to-year variation occurs in the interaction structure of a community at the same site in the absence of significant disturbance such as death of a canopy tree?

CONCLUSIONS

This final chapter has been an extended argument for the analysis of interactions between species within the context of communities in which they live. The major points are these:

1. The evolutionary unit of interaction is the group of species within which selection acting on one of the pairs of species significantly affects selection on the other species. The evolutionary unit of interaction often involves more than two species, and interactions between species must be studied in this context.

2. The number of species involved in an interaction can grow over evolutionary time through both cospeciation and through the collection of unrelated species. Mutualisms that produce new resources (fruits, nectar) seem to be the interactions most likely to grow rapidly through the collection of unrelated species.

3. Mutualisms can form major links among component communities and these interactions can be of major importance in the organization of communities.

4. The direct interactions among species can ripple in their effects through at least some of the remaining species in communities. We still know little about the indirect effects of interactions.

5. Interactions between species vary in their probability of occur-

rence with disturbances regimes and environmental gradients. Research on the patch dynamics of interactions should focus on where different kinds of interactions occur within the context of disturbance regimes and environmental gradients, and how interactions change among species at a particular site over time. These analyses can help to specify the ecological conditions most likely to lead to coevolution between species and are, therefore, the basis for understanding the evolution of interactions and the ecology of coevolution.

REFERENCES

Abrams, P. A. 1980. Resource partitioning and interspecific competition in a tropical hermit crab community. *Oecologia* **46**: 365–379.

Addicott, J. F. 1978. Competition for mutualists: aphids and ants. *Can. J. Zool.* **56**: 2093–2096.

Addicott, J. F. 1979. A multispecies aphid-ant association: density dependence and species-specific effects. *Can. J. Zool.* **57**: 558–569.

Addicott, J. F. 1981. Stability properties of 2-species models of mutualism: simulation studies. *Oecologia* **49**: 42–49.

Ahmadjian, V. 1982. The nature of lichens. *Nat. Hist.* **91**(3): 31–36.

Aker, C. L., and D. Udovic. 1981. Oviposition and pollination behavior of the yucca moth, *Tegeticula maculata* (Lepidoptera:Prodoxidae) and its relation to the reproductive biology of *Yucca whipplei* (Agavaceae) *Oecologia* **49**: 96–101.

Alexander, R. D. 1974. The evolution of social behavior. *Annu. Rev. Ecol. Syst.* **5**: 325–383.

Allison, M. J., and H. M. Cook. 1981. Oxalate degradation by microbes of the bowel of herbivores: the effect of dietary oxalate. *Science* **212**: 675–676.

Alloway, T. M. 1972. Learning and memory in insects. *Annu. Rev. Entomol.* **17**: 43–56.

Alstad, D. N., G. F. Edmunds, Jr., and S. C. Johnson. 1980. Host adaptation, sex ratio, and flight activity in male black pineleaf scale. *Ann. Entomol. Soc. Am.* **73**: 665–667.

Anderson, G. R. V., A. H. Ehrlich, P. R. Ehrlich, J. D. Roughgarden, B. C. Russell, and F. H. Talbot. 1981. The community structure of coral reef fishes. *Am. Nat.* **17**: 476–495.

Arnold, S. J. 1977. Polymorphism and geographic variation in the feeding behavior of the garter snake *Thamnophis elegans*. *Science* **197**: 676–678.

Arnold, S. J. 1980. The microevolution of feeding behavior. Pages 409–453 in A. Kamil and T. Sargent, Eds. *Foraging behavior: ecological, ethological, and psychological approaches.* Garland, New York.

Arnold, S. J. 1981a. Behavioral variation in natural populations. I. Phenotypic, genetic and environmental correlations between chemoreceptive responses to prey in the garter snake, *Thamnophis elegans*. *Evolution* **35**: 489–509.

Arnold, S. J. 1981b. Behavioral variation in natural populations. II. The inheritance of a feeding response in crosses between geographic races of the garter snake, *Thamnophis elegans. Evolution* **35**: 510–515.

Askew, R. R. 1971. *Parasitic insects.* American Elsevier, New York.

Atsatt, P. R. 1981a. Ant-dependent food plant selection by the mistletoe butterfly *Ogyris amaryllis* (Lycaenidae). *Oecologia* **48**: 60–63.

Atsatt, P. R. 1981b. Lycaenid butterflies and ants: selection for enemy-free space. *Am. Nat.* **118**: 638–654.

Augsberger, C. K. 1981. Reproductive synchrony of a tropical shrub: experimental studies on effects of pollinators and seed predators on *Hybanthus prunifolius* (Violaceae). *Ecology* **62**: 775–788.

Axelrod, D. I. 1970. Mesozoic paleogeography and early angiosperm history. *Bot. Rev.* **36**: 277–319.

Bailey, I. W. 1922. The anatomy of certain plants from the Belgian Congo, with special reference to myrmecophytism. *Bull. Am. Mus. Nat. Hist.* **45**: 585–621.

Baker, H. G., and I. Baker. 1979. Starch in angiosperm pollen grains and its evolutionary significance. *Am. J. Bot.* **66**: 591–600.

Balda, R. P. 1980. Are seed caching systems co-evolved? *Proc. 17th Int. Ornith. Congr.*, pp. 1185–1191.

Barnes, O. L. 1965. Further tests of the effects of food plants on the migratory grasshopper. *J. Econ. Entomol.* **58**: 475–479.

Bartz, S. H. 1979. Evolution of eusociality in termites. *Proc. Natl. Acad. Sci. USA 76*, pp. 5764–5768.

Batzli, G. O., R. G. White, S. F. MacLean, Jr., F. A. Pitelka, and D. B. Collier. 1980. The herbivore-based trophic system. Pages 335–410 in J. Brown, P. C. Miller, L. L. Tieszen, and F. L. Bunnell, eds. *An arctic ecosystem: the coastal tundra at Barrow, Alaska.* Dowden, Hutchinson, & Ross, Stroudsburg, PA.

Bawa, K. S. 1980. Evolution of dioecy in flowering plants. *Annu. Rev. Ecol. Syst.* **11**: 15–39.

Bazzaz, F. A., and J. L. Harper. 1977. Demographic analysis of the growth of *Linum usitatissimum. New Phytol.* **78**: 193–208.

Beattie, A. J., and D. C. Culver. 1981. The guild of myrmecochores in the herbaceous flora of West Virginia forests. Ecology **62**: 107–115.

Becker, P. 1975. Island colonization by carnivorous and herbivorous Coleoptera. *J. Anim. Ecol.* **44**: 893–906.

Bell, C. R. 1971. Breeding systems and floral biology of the Umbelliferae or Evidence for specialization in unspecialized flowers. Pages 93–107 in V. H. Heywood, Ed. *The biology and chemistry of the Umbelliferae.* Academic, New York.

Belt, T. 1874. *The naturalist in Nicaragua.* Murray, London.

Benson, W. W. 1978. Resource partitioning in passion vine butterflies. *Evolution* **32**: 493–518.

Benson, W. W., K. S. Brown, Jr., and L. E. Gilbert, 1975. Coevolution of plants and herbivores: passion flower butterflies. *Evolution* **29**: 659–680.

Bentley, B. L. 1976. Plants bearing extrafloral nectaries and the associated ant community: interhabitat differences in the reduction of herbivore damage. *Ecology* **57**: 815–820.

Bentley, B. L. 1977a. Extrafloral nectaries and protection by pugnacious bodyguards. *Annu. Rev. Ecol. Syst.* **8**: 407–427.

Bentley, B. L. 1977b. The protective function of ants visiting the extrafloral nectaries of *Bixa orellana* (Bixaceae). *J. Ecology* **65:** 27–38.

Benzing, D. H., and A. Renfrow. 1974. The mineral nutrition of Bromeliaceae. *Bot. Gaz.* **135:** 281–288.

Bequaert, J. 1922. Ants in their diverse relations to the plant world. *Bull. Am. Mus. Nat. Hist.* **45:** 333–585.

Berg, R. Y. 1958. Seed dispersal, morphology, and phylogeny of *Trillium*. *Skr. Norske Vidensk.-Akad. Oslo. Mat.-Naturv. Klasse* **1958**(1): 1–36.

Berg, R. Y. 1975. Myrmecochorous plants in Australia and their dispersal by ants. *Aust. J. Bot.* **23:** 475–508.

Bernays, E. A. 1978. Tannins: an alternative viewpoint. *Entomol. Exp. and Appl.* **24:** 44–53.

Bertness, M. D. 1981. Competitive dynamics of a tropical hermit crab assemblage. *Ecology* **62:** 751–761.

Bossema, I. 1979. Jays and oaks: an eco-ethological study of a symbiosis. *Behaviour* **70:** 1–117.

Brambell, M. R. 1976. The giant panda *(Ailuropoda melanoleuca)*. *Trans. Zool. Soc. Lond.* **33:** 85–92.

Breedlove, D. E., and P. R. Ehrlich. 1972. Coevolution: patterns of legume predation by a lycaenid butterfly. *Oecologia* **10:** 99–104.

Brittain, R. J., and E. H. Davidson. 1971. Repetitive and non-repetitive DNA sequence and a speculation on the origins of evolutionary novelties. *Quart. Rev. Biol.* **46:** 111–138.

Brooks, D. R. 1979a. Testing the context and extent of host–parasite coevolution. *Syst. Zool.* **28:** 299–307.

Brooks, D. R. 1979b. Testing hypotheses of evolutionary relationships among parasites: the digeneans of crocodileans. *Am. Zool.* **19:** 1225–1238.

Brooks, D. R. 1981. Review of P. W. Price, evolutionary biology of parasites. *Syst. Zool.* **30:** 104–107.

Brower, J. V. Z. 1958a. Experimental studies of mimicry in some North American butterflies. Part I. The monarch, *Danaus plexippus* and viceroy, *Limenitis archippus archippus*. *Evolution* **12:** 32–47.

Brower, J. V. Z. 1958b. Experimental studies of mimicry in some North American butterflies. Part II. *Battus philenor* and *Papilio troilus, P. polyxenes* and *P. glaucus*. *Evolution* **12:** 123–136.

Brower, L. P. 1958. Bird predation and foodplant specificity in closely related procryptic insects. *Am. Nat.* **92:** 183–187.

Brower, L. P., J. V. Z. Brower, F. G. Stiles, H. J. Croze, and A. S. Hower. 1964. Mimicry: differential advantage of color patterns in the natural environment. *Science* **144:** 183–185.

Brown, J. H., and A. Kodric-Brown. 1979. Convergence, competition, and mimicry in a temperate community of hummingbird-pollinated flowers. *Ecology* **60:** 1022–1035.

Brown, K. S. 1976. Geographical patterns of evolution in neotropical forest Lepidoptera (Nymphalidae: Ithomiinae and Nymphalinae-Heliconiini). Pages 118–160 in H. Descimon, Ed. *Biogéographie et évolution en Amerique tropicale*. Publications du Laboratoire de Zoologie de L'Ecole Normale Supérieure No. 9.

Brown, K. S., P. M. Sheppard, and J. F. G. Turner. 1974. Quaternary refugia in tropical America: evidence from race formation in *Heliconius* butterflies. *Proc. R. Soc. Lond. B* **187:** 369–378.

Brues, C. T. 1924. The specificity of foodplants in the evolution of phytophagous insects. *Am. Nat.* **58:** 127–144.

Burnett, J. H. 1975. *Mycogenetics.* Wiley, New York.

Bush, G. L. 1975. Modes of animal speciation. *Annu. Rev. Ecol. Syst.* **6:** 339–364.

Carlquist, S. 1974. *Island biology.* Columbia University Press, New York.

Carroll, C. R.,.and D. H. Janzen. 1973. Ecology of foraging by ants *Annu. Rev. Ecol. Syst.* **4:** 231–257.

Case, T. J. 1979. Character displacement and coevolution in some *Cnemidophorus* lizards. *Fortschr. Zool.* **25:** 235–282.

Case, T. J., and M. E. Gilpin. 1974. Interference competition and niche theory. *Proc. Natl. Acad. Sci. USA 71,* pp. 3073–3077.

Cates, R. G. 1980. Feeding patterns of monophagous, oligophagous, and polyphagous insect herbivores: the effect of resource abundance and plant chemistry. *Oecologia* **46:** 22–31.

Cates, R. G., and D. F. Rhoades. 1977. Patterns in the production of antiherbivore chemical defenses in plant communities. *Biochem. Syst. Ecol.* **5:** 185–193.

Chandler, G. E., and J. W. Anderson. 1976. Studies on the nutrition and growth of *Drosera* species with reference to the carnivorous habit. *New Phytol.* **76:** 129–141.

Charles-Dominique, P. 1977. *Ecology and behaviour of nocturnal primates. Prosimians of Equatorial West Africa.* Duckworths, London.

Cherry, L. M., S. M. Case, and A. C. Wilson. 1978. Frog perspective on the morphological difference between humans and chimps. *Science* **200:** 209–211.

Chew, F. S. 1975. Coevolution of pierid butterflies and their cruciferous foodplants. I. The relative quality of available resources. *Oecologia* **20:** 117–127.

Chew, F. S. 1977. Coevolution of pierid butterflies and their cruciferous foodplants. II. The distribution of eggs on potential food-plants. *Evolution* **31:** 568–579.

Chew, F. S. 1980. Foodplant preferences of *Pieris* caterpillars (Lepidoptera). *Oecologia* **46:** 347–353.

Clarke, B. 1976. The ecological genetics of host–parasite relationships. Pages 87–103 in A. E. R. Taylor and R. Muller, Eds. *Genetic aspects of host parasite relationships. Symp. Br. Soc. Parasitol.* No. 14.

Clarke, J. F. G. 1952. Host relationships of moths of the genera *Depressaria* and *Agonopterix,* with descriptions of new species. *Smithsonian Misc. Coll.* **117**(7): 1–20.

Clifford, H. T. 1962. Insect pollinators of *Plantago lanceolata* L. *Nature* **193:** 196.

Coley, P. D. 1980. Effects of leaf age and plant life history patterns on herbivory. *Nature* **284:** 545–546.

Collins, L., and M. Roberts. 1978. Arboreal folivores in captivity—maintenance of a delicate minority. Pages 5–12 in G. G. Montgomery, Ed. *The ecology of arboreal folivores.* Smithsonian Institution Press, Washington, DC.

Colwell, R. K., and E. R. Fuentes. 1975. Experimental studies of the niche. *Annu. Rev. Ecol. Syst.* **6:** 281–310.

Congdon, J. D., L. J. Vitt, and W. W. King. 1974. Geckos: Adaptive significance and energetics of tail autotomy. *Science* **184:** 1379–1380.

Connell, J. H. 1961a. The influence of interspecific competition and other factors on the distribution of the barnacle *Chthamalus stellatus. Ecology* **42:** 710–723.

Connell, J. H. 1961b. Effect of competition predation by *Thais lapillus,* and other factors on natural populations of the barnacle *Balanus balanoides. Ecol. Monogr.* **31:** 61–104.

Connell, J. H. 1973. Population ecology of reef-building corals. Pages 205–245 in O. A. Jones and R. Endean, Eds. *Biology and geology of coral reefs. Vol. II: Biology I.* Academic, New York.

Connell, J. H. 1975. Some mechanisms producing structure in natural communities: a model and evidence from field experiments. Pages 460–490 in M. L. Cody and J. M. Diamond, Eds. *Ecology and evolution of communities.* Harvard University Press, Cambridge, MA.

Connell, J. H. 1978. Diversity in tropical rain forests and coral reefs. *Science* **199**: 1302–1310.

Connell, J. H. 1980. Diversity and the coevolution of competitors, or the ghost of competition past. *Oikos* **35**: 131–138.

Connor, E. F., and D. Simberloff. 1979. The assembly of species communities: chance or competition. *Ecology* **60**: 1132–1140.

Cornell, H. V., and D. Pimentel. 1978. Switching in the parasitoid *Nasonia vitripennis* and its effects on host competition. *Ecology* **59**: 297–308.

Crepet, W. L. 1979. Insect pollination: a paleontological perspective. *Bioscience* **29**: 102–108.

Crome, F. H. J. 1975. The ecology of fruit pigeons in tropical northern Queensland. *Aust. Wildl. Res.* **2**: 155–185.

Cruden, R. W. 1972a. Information on chemistry and pollination biology relevant to the systematics of *Nemophila menziesii* (Hydrophyllaceae). *Mandrono* **21**: 505–515.

Cruden, R. W. 1972b. Pollination biology of *Nemophila menziesii* (Hydrophyllaceae) with comments on the evolution of oligolectic bees. *Evolution* **26**: 373–389.

Culver, D. C., and A. J. Beattie. 1978. Myrmecochory in *Viola:* dynamics of seed–ant interactions in some West Virginia species. *J. Ecol.* **66**: 53–72.

Culver, D. C., and A. J. Beattie. 1980. The fate of *Viola* seeds dispersed by ants. *Amer. J. Bot.* **67**: 710–714.

Dadd, R. H. 1973. Insect nutrition: current developments and metabolic implications. *Annu. Rev. Entomol.* **18**: 381–420.

Darwin, C. 1859. *On the origin of species by means of natural selection.* Reprinted 1964. Harvard University Press, Cambridge, MA.

Davidson, D. W., and S. R. Morton. 1981. Competition for dispersal in ant-dispersed plants. *Science* **213**: 1259–1261.

Davis, D. D. 1964. The giant panda. A morphological study of evolutionary mechanisms. *Fieldiana: Zool. Mem.* **3**: 1–339.

Davis, D. R. 1967. A revision of the moths of the subfamily Prodoxinae (Lepidoptera:Incurvariidae) *Bull. US Nat. Mus.* **255**: 1–170.

Davis, J. 1973. Habitat preferences and competition of wintering juncos and golden-crowned sparrows. *Ecology* **54**: 174–180.

Day, P. R. 1974. *Genetics of host–parasite interactions.* Freeman, San Francisco.

Dayton, P. K. 1971. Competition, disturbance, and community organization: the provision and subsequent utilization of space in a rocky intertidal community. *Ecol. Monogr.* **41**: 351–389.

Dayton, P. K. 1973. Two cases of resource partitioning in an intertidal community: making the right prediction for the wrong reason. *Am. Nat.* **107**: 662–670.

Denslow, J. S. 1980a. Patterns of plant species diversity during succession under different disturbance regimes. *Oecologia* **46**: 18–21.

Denslow, J. S. 1980b. Gap partitioning among tropical rainforest trees. *Biotropica* **12** (Supplement): 47–55.

Dethier, V. G. 1954. Evolution of feeding preferences in phytophagous insects. *Evolution* **8:** 33–54.

Dethier, V. G. 1980. Food-aversion in two polyphagous caterpillars, *Diacrisia virginica* and *Estigmene congrua. Physiol. Entomol.* **5:** 321–325.

Dhondt, A. A., and R. Eyckerman. 1980. Competition between the Great Tit and the Blue Tit outside the breeding season in field experiments. *Ecology* **61:** 1291–1296.

Dindal, D. L. 1975. Symbiosis: nomenclature and proposed classification, *Biologist* **57:** 129–142.

Dobzhansky, T. 1970. *Genetics of the evolutionary process.* Columbia University Press, New York.

Dolinger, P. M., P. R. Ehrlich, W. L. Finch, and D. E. Breedlove. 1973. Alkaloid and predation patterns in Colorado lupine populations. *Oecologia* **13:** 191–204.

Dressler, R. L. 1968. Pollination in euglossine bees. *Evolution* **22:** 202–210.

Dunham, A. E. 1980. An experimental study of interspecific competition between the iquanid lizards *Sceloporus merriami* and *Urosaurus ornatus. Ecol. Mongr.* **50:** 309–330.

Dunham, A. E., G. R. Smith, and J. N. Taylor. 1979. Evidence for ecological character displacement in western American catostomid fishes. *Evolution* **33:** 877–896.

Dunlap-Pianka, H., C. L. Boggs, and L. E. Gilbert. 1977. Ovarian dynamics in heliconiine butterflies: programmed senescence versus eternal youth. *Science* **197:** 487–490.

Dyer, M. I. 1975. The effects of red-winged blackbirds *(Agelaius phoeniceus* L.) on biomass production of corn grains *(Zea mays* L.) *J. Appl. Ecol.* **12:** 719–726.

Ebbers, B. C., and E. M. Barrows. 1980. Individual ants specialize on particular aphid herds (Hymenoptera: Formicidae; Homoptera:Aphididae). *Proc. Entomol. Soc. Wash.* 82, pp. 405–407.

Edmunds, G. G., Jr. and D. N. Alstad. 1978. Coevolution in insect herbivores and conifers. *Science* **199:** 941–945.

Edson, K. M., S. B. Vinson, D. B. Stoltz, and M. D. Summers. 1981. Virus in a parasitoid wasp: suppression of the cellular immune response in the parasitoid's host. *Science* **211:** 582–583.

Egerton, F. N. 1973. Changing concepts in the balance of nature. *Quart. Rev. Biol.* **48:** 322–350.

Ehrlich, P. R., and P. H. Raven. 1964. Butterflies and plants: a study in coevolution. *Evolution* **18:** 586–608.

Eickwort, G. C. and H. S. Ginsberg. 1980. Foraging and mating behavior in Apoidea. *Annu. Rev. Entomol.* **25:** 421–426.

Eldredge, N., and S. J. Gould. 1972. Punctuated equilibria: an alternative to phyletic gradualism. Pages 82–115 in T. J. M. Schopf, Ed. *Models in paleobiology.* Freeman, Cooper & Co., San Francisco.

Elias, T. S. 1980. Foliar nectaries of unusual structure in *Leonardoxa africana* (Leguminosae), an African obligate myrmecophyte. *Am. J. Bot.* **67:** 423–425.

Elton, C. S. 1966. *The pattern of animal communities.* Methuen, London.

Endler, J. A. 1977. *Geographic variation, speciation and clines.* Princeton University Press, Princeton, NJ.

Evans, H. E. 1977. Extrinsic versus intrinsic factors in the evolution of insect sociality. *BioScience* **27**: 613–617.

Feeny, P. 1975. Biochemical coevolution between plants and their insect herbivores. Pages 3–19 in L. E. Gilbert and P. H. Raven, Eds. *Coevolution of animals and plants.* University of Texas Press, Austin.

Feeny, P. 1976. Plant apparency and chemical defense. *Rec. Adv. Phytochem.* **10**: 1–40.

Feinsinger, P., and R. K. Colwell. 1978. Community organization among neotropical nectar-feeding birds. *Am. Zool.* **18**: 779–795.

Fenchel, T. 1975. Character displacement and coexistence in mud snails. *Oecologia* **20**: 19–32.

Fenchel, T., and L. H. Kafoed. 1976. Evidence for exploitative interspecific competition in mud snails (Hydrobiidae). *Oikos* **27**: 367–376.

Flor, H. H. 1942. Inheritance of pathogenicity in *Melampsora lini. Phytopathology* **32**: 653–659.

Flor, H. H. 1955. Host–parasite interaction in flax rust—its genetics and other implications. *Phytopathology* **45**: 680–685.

Flor, H. H. 1971. Current status of the gene-for-gene concept. *Annu. Rev. Phytopath.* **9**: 275–295.

Forcella, F. 1980. Cone predation by pinyon cone beetle *(Conopthorus edulis;* Scolytidae): dependence on frequency and magnitude of cone production. *Am. Nat.* **116**: 594–598.

Foster, M. S. 1978. Total frugivory in tropical passerines: a reappraisal. *Trop. Ecol.* **19**: 131–154.

Fox, L. R. 1981. Defense and dynamics in plant–herbivore systems. *Am. Zool.* **24**: 853–864.

Fox, L. R., and B. J. Macauley. 1977. Insect grazing on *Eucalyptus* in response to variation in leaf tannins and nitrogen. *Oecologia* **29**: 145–162.

Fox, L. R., and P. A. Morrow. 1981. Specialization: species property or local phenomenon? *Science* **211**: 887–893.

Fraenkel, G. S. 1959. The raison d'être of secondary plant substances. *Science* **129**: 1466–1470.

Fraenkel, G. S. 1969. Evaluations of our thoughts on secondary plant substances. *Entomol. Exp. Appl.* **12**: 473–486.

Frankie, G. W., P. A. Opler, and K. S. Bawa. 1976. Foraging behaviour of solitary bees: implications for outcrossing of a neotropical forest tree species. *J. Ecol.* **64**: 1049–1057.

Freeland, W. J., and D. H. Hanzen. 1974. Strategies in herbivory by mammals: the role of plant secondary compounds. *Am. Nat.* **108**: 269–289.

Frith, H. J., F. H. J. Crome, and T. O. Wolfe. 1976. Food of fruit-pigeons in New Guinea. *Emu* **76**: 49–58.

Frost, P. G. H. 1980. Fruit–frugivore interactions in a South African coastal dune forest. *Proc. 17th Int. Ornith. Congr.,* pp. 1179–1184.

Futuyma, D. J. 1976. Food plant specialization and environmental predictability in Lepidoptera. *Am. Nat.* **110**: 285–292.

Galil, J. 1973. Topocentric and ethodynamic pollination. Pages 85–100 in N.M.B. Brantjes, Ed. *Pollination and dispersal.* University of Nijmegen, Nijmegen.

Gallun, R. L. 1977. The genetic basis of Hessian fly epidemics. *Ann. N. Y. Acad. Sci.* **287**: 223–229.

Gallun, R. L., and G. S. Khush. 1980. Genetic factors affecting expression and stability of resistance. Pages 64–85 in F. G. Maxwell and P. R. Jennings, Eds. *Breeding plants resistant to insects.* Wiley–Interscience, New York.

Gallun, R. L., K. J. Starks, and W. D. Guthrie. 1975. Plant resistance to insects attacking cereals. *Annu. Rev. Entomol.* **20:** 337–357.

Gause, G. F. 1932. *The struggle for existence.* Williams and Wilkins, Baltimore.

Gauthier-Pilters, H., and A. I. Dagg. 1981. *The camel: its evolution, ecology, behavior, and relationship to man.* University of Chicago Press, Chicago.

Gilbert, L. E. 1972. Pollen feeding and the reproductive biology of *Heliconius* butterflies. *Proc. Natl. Acad. Sci. USA* 69, pp. 1403–1407.

Gilbert, L. E. 1975. Ecological consequences of a coevolved mutualism between butterflies and plants. Pages 210–240 in L. E. Gilbert and P. H. Raven, Eds. *Coevolution of animals and plants.* University of Texas Press, Austin.

Gilbert, L. E. 1977. The role of insect–plant coevolution in the organization of ecosystems. Pages 399–413 in V. Labyrie, Ed. *Comportement des Insectes et Milieu Trophique.* C.N.R.S., Paris.

Gilbert, L. E. 1979. Development of theory in the analysis of insect–plant interactions. Pages 117–154 in D. J. Horn, R. D. Mitchell, and G. R. Stairs, Eds. *Analysis of ecological systems.* Ohio State University Press, Columbus.

Gilbert, L. E. 1980. Food web organization and the conservation of neotropical diversity. Pages 11–33 in M. E. Soule and B. A. Wilcox, Eds. *Conservation biology: an evolutionary–ecological perspective.* Sinauer Associates, Sunderland, MA.

Gilbert, L. E. and P. H. Raven, Eds. 1975. *Coevolution of animals and plants.* University of Texas Press, Austin.

Gilbert, L. E. and M. C. Singer. 1975. Butterfly ecology. *Annu. Rev. Ecol. Syst.* **6:** 365–397.

Gill, Douglas E. 1974. Intrinsic rate of increase, saturation density, and competitive ability II. The evolution of competitive ability. *Am. Nat.* **108:** 103–116.

Glander, K. E. 1978. Howler monkey feeding behavior and plant secondary compounds: a study of strategies. Pages 561–574 in G. G. Montgomery, Ed. *The ecology of arboreal folivores.* Smithsonian Institution Press, Washington, DC.

Glynn, P. W. 1980. Defense by symbiotic crustacea of host corals elicited by chemical cues from predator. *Oecologia* **47:** 287–290.

Goh, G. B. 1979. Stability in models of mutualism. *Am. Nat.* **113:** 261–275.

Gorlick, D. L., P. D. Atkins, and G. S. Losey, Jr. 1978. Cleaning stations as water holes, garbage dumps, and sites for the evolution of reciprocal altruism? *Am. Nat.* **112:** 341–353.

Gould, S. J. 1977. *Ontogeny and phylogeny.* Harvard University Press, Cambridge, MA.

Gould, S. J. 1980. The evolutionary biology of constraint. *Daedalus* **109:** 39–52.

Gould, S. J., and N. Eldredge. 1977. Punctuated equilibria: the tempo and mode of evolution reconsidered. *Paleobiology* **3:** 115–151.

Grace, J. B., and R. G. Wetzel. 1981. Habitat partitioning and competitive displacement in cattails *(Typha):* experimental field studies. *Am. Nat.* **118:** 463–474.

Grant, K. A. 1966. An hypothesis concerning the prevalence of red coloration in California hummingbird flowers. *Am. Nat.* **100:** 85–98.

Grant, K. A., and V. Grant. 1968. *Hummingbirds and their flowers.* Columbia University Press, New York.

Grant, P. R. 1972. Convergent and divergent character displacement. *Biol. J. Linn. Soc.* **4:** 39–68.

Grant, P. R. 1975. The classical case of character displacement. *Evol. Biol.* **8:** 237–337.

Grant, P. R., and I. Abbott. 1980. Interspecific competition, island biogeography and null hypotheses. *Evolution* **34:** 332–341.

Grant, V. 1950. The flower constancy of bees. *Bot. Rev.* **16:** 379–398.

Grant, V. 1971. *Plant speciation.* Columbia University Press, New York.

Grubb, P. J. 1977. The maintenance of species richness in plant communities: the importance of the regeneration niche. *Biol. Rev.* **52:** 107–145.

Hairston, N. G. 1951. Interspecies competition and its probable influence upon the vertical distribution of Appalachian salamanders of the genus *Plethodon. Ecology* **32:** 266–274.

Hairston, N. G. 1973. Ecology, selection, and systematics. *Breviora* **414:** 1–21.

Hairston, N. G. 1980a. The experimental test of an analysis of field distributions: competition in terrestrial salamanders. *Ecology* **61:** 817–826.

Hairston, N. G. 1980b. Evolution under interspecific competition: field experiments on terrestrial salamanders. *Evolution* **34:** 409–420.

Hairston, N. G. 1980c. Species packing in the salamander genus *Desmognathus:* what are the interspecific interactions involved? *Am. Nat.* **115:** 354–366.

Hairston, N. G. 1981. An experimental test of a guild: salamander competition. *Ecology* **62:** 65–72.

Hamilton, W. D. 1964. The genetical evolution of social behaviour. I, II. *J. Theor. Biol.* **7:** 1–16; **7:** 17–52.

Hamilton, W. D. 1972. Altruism and related phenomena, mainly in social insects. *Annu. Rev. Ecol. Syst.* **3:** 193–232.

Handel, S. N. 1978. The competitive relationship of three woodland sedges and its bearing on the evolution of ant-dispersal of *Carex pedunculata. Evolution* **32:** 151–163.

Harger, J. R. E. 1972. Competitive co-existence: maintenance of interacting associations of the sea mussels *Mytilus edulis* and *Mytilus californianus. Veliger* **14:** 387–410.

Harger, J. R. E. 1972b. Competitive coexistence among intertidal invertebrates. *Am. Sci.* **60:** 600–607.

Harper, J. L. 1977. *Population biology of plants.* Academic, New York.

Hartshorn, G. S. 1980. Neotropical forest dynamics. *Biotropica* **12** (supplement): 23–30.

Haukioja, E., and P. Niemelä. 1979. Birch leaves as a resource for herbivores: seasonal occurrence of increased resistance in foliage after mechanical damage of adjacent leaves. *Oecologia* **39:** 151–159.

Haven, S. B. 1973. Competition for food between the intertidal gastropods *Acmaea scabra* and *Acmaea digitalis. Ecology* **54:** 143–151.

Heatwole, H., and R. Levins. 1973. Biogeography of a Puerto Rican bank: species turnover on a small cay, Cayo Ahogado. *Ecology* **54:** 1042–1055.

Heinrich, B. 1975. Bee flowers: a hypothesis on flower variety and blooming times. *Evolution* **29:** 325–334.

Heinrich, B. 1976. The foraging specializations of individual bumblebees. *Ecol. Monogr.* **46:** 105–128.

Heinrich, B. 1979. "Majoring" and "minoring" by foraging bumblebees, *Bombus vagans:* an experimental analysis. *Ecology* **60:** 245–255.

Heithaus, E. R., D. C. Culver, and A. J. Beattie. 1980. Models of some ant–plant mutualisms. *Am. Nat.* **116**: 347–361.

Hendrickson, J. A., Jr. 1981. Community-wide character displacement reexamined. *Evolution* **35**: 794–800.

Hendrix, S. D. 1980. An evolutionary and ecological perspective of the insect fauna of ferns. *Am. Nat.* **115**: 171–196.

Hespenheide, H. A. 1979. Are there fewer parasitoids in the tropics? *Am. Nat.* **113**: 766–769.

Hinton, H. E. 1951. Hyrmecophilous Lycaenidae and other Lepidoptera—a summary. *Trans. S. Lond. Entomol. Nat. Hist. Soc. 1949–1950:* 111–175.

Hixon, M. A. 1980. Competitive interactions between California reef fishes of the genus *Embiotoca. Ecology* **61**: 918–931.

Hobson, E. S. 1969. Comments on generalizations about cleaning symbiosis in fishes. *Pac. Sci.* **23**: 35–39.

Hocking, B. 1970. Insect associations with swollen thorn acacias. *Trans. R. Entomol. Soc. Lond.* **122**: 211–255.

Hocking, B. 1975. Ant–plant mutualism: evolution and energy. Pages 78–90 in L. E. Gilbert and P. H. Raven, Eds. *Coevolution of animals and plants.* University of Texas Press, Austin.

Hodges, R. W. 1974. *Gelechioidea:Oecophoridae.* Fasc. 6.2 in R. B. Dominick, et al., Eds. *The moths of America north of Mexico.* Classey Limited, London.

Holling, C. S. 1959. The components of predation as revealed by a study of small mammal predation of the European pine sawfly. *Can. Entomol.* **91**: 293–320.

Holling, C. S. 1961. Principles of insect predation. *Annu. Rev. Ecol. Syst.* **6**: 163–182.

Holling, C. S. 1980. Implications of parasitism. Review of P. W. Price, evolutionary biology of parasites. *Science* **210**: 1240–1241.

Holmes, J. C. 1973. Site selection by parasitic helminths: interspecific interactions, site segregation, and their importance to the development of helminth communities. *Can. J. Zool.* **51**: 333–347.

Horvitz, C. C., and A. J. Beattie. 1980. Ant dispersal of *Calathea* (Maranthaceae) seeds by carnivorous ponerines (Formicidae) in a tropical rain forest. *Am. J. Bot.* **67**: 321–326.

Houk, E. J., and G. W. Griffiths. 1980. Intracellular symbiotes of the Homoptera. *Annu. Rev. Entomol.* **25**: 161–187.

Howe, H. F. 1977. Bird activity and seed dispersal of a tropical wet forest tree. *Ecology* **58**: 539–550.

Howe, H. F. 1979. Fear and frugivory. *Am. Nat.* **114**: 925–931.

Howe, H. F. 1981. Dispersal of a neotropical nutmeg *(Virola sebifera)* by birds. *Auk* **98**: 88–98.

Howe, H. F., and D. De Steven. 1979. Fruit production, migrant bird visitation, and seed dispersal of *Guarea glabra* in Panama. Oecologia **39**: 185–196.

Howe, H. F., and G. F. Estabrook. 1977. On intraspecific competition for avian dispersers in tropical trees. *Am. Nat.* **111**: 817–832.

Howe, H. F., and G. A. Vande Kerckhove. 1979. Fecundity and seed dispersal of a tropical tree. *Ecology* **60**: 180–189.

Howe, H. F., and G. A. Vande Kerckhove. 1980. Nutmeg dispersal by tropical birds. *Science* **210**: 925–927.

Hubbell, S. P. 1979. Tree dispersion, abundance, and diversity in a tropical dry forest. *Science* **203:** 1299–1309.

Huey, R. B., E. R. Pianka, M. E. Egan, and L. W. Coons. 1974. Ecological shifts in sympatry: Kalahari fossorial lizards *(Typhlosaurus)*. *Ecology* **55:** 304–316.

Hungate, R. E. 1975. The rumen microbial ecosystem. *Annu. Rev. Ecol. Syst.* **6:** 39–66.

Huston, M. 1979. A general hypothesis of species diversity. *Am. Nat.* **113:**81–101.

Hutchinson, G. E. 1959. Homage to Santa Rosalie or Why are there so many kinds of animals? *Am. Nat.* **93:**145–159.

Hutchinson, G. E. 1961. The paradox of the plankton. *Am. Nat.* **95:**137–145.

Hutchinson, G. E. 1965. *The ecological theater and the evolutionary play.* Yale University Press, New Haven, CT.

Huxley, C. R. 1978. The ant-plants *Myrmecodia* and *Hydnophytum* (Rubiaceae) and the relationships between their morphology, ant occupants, physiology and ecology. *New Phytol.* **80:**231–268.

Huxley, C. 1980. Symbiosis between ants and epiphytes. *Biol. Rev.* **55:**321–340.

Inger, R. F., and R. K. Colwell. 1977. Organization of contiguous communities of amphibious and reptiles in Thailand. *Ecol. Monogr.* **47:**229–253.

Inouye, D. W. 1978. Resource partitioning in bumblebees: experimental studies of foraging behavior. *Ecology* **59:**672–678.

Inouye, D. W. 1980. The terminology of floral larceny. *Ecology* **61:**1251–1253.

Inouye, D. W., and O. R. Taylor, Jr. 1979. A temperate region plant–ant–seed predator system: consequences of extrafloral nectar secretion by *Helianthella guinquenervis*. *Ecology* **60:**1–7.

Itzkowitz, M. 1979. The feeding strategies of a facultative cleaner fish *Thalassama bifasciatum* (Pisces:Labridae). *J. Zol.* **187:**403–413.

Janos, D. P. 1980. Mycorrhizae influence tropical succession. *Biotropica* **12** (Supplement): 56–64.

Janzen, D. H. 1966. Coevolution of mutualism between ants and acacias in Central America. *Evolution* **20:**249–275.

Janzen, D. H. 1967. Interaction of the bull's-horn acacia *(Acacia cornigera* L.) with its ant inhabitant *(Pseudomyrmex ferruginea* F. Smith) in eastern Mexico. *Univ. Kansas Sci. Bull.* **47:**315–558.

Janzen, D. H. 1969. Birds and the ant X acacia interaction in Central America, with notes on birds and other myrmecophytes. *Condor* **71:**240–256.

Janzen, D. H. 1971a. Seed predation by animals. *Annu. Rev. Ecol. Syst.* **2:**465–492.

Janzen, D. H. 1971b. Euglossine bees as long-distance pollinators of tropical plants. *Science* **171:**203–205.

Janzen, D. H. 1973. Host plants as islands. II. Competition in evolutionary and contemporary time. *Am. Nat.* **107:**786–790.

Janzen, D. H. 1974a. Tropical blackwater rivers, animals, and mast fruiting by the Dipterocarpaceae. *Biotropica* **6:**69–103.

Janzen, D. H. 1974b. Swollen-thorn acacias of Central America. *Smithsonian Contrib. Bot.* **13:**1–131.

Janzen, D. H. 1974c. Epiphytic myrmecophytes in Sarwak: mutualism through the feeding of plants by ants. *Biotropica* **6:**237–259.

Janzen, D. H. 1975a. *Pseudomyrmex nigropilosa:* a parasite of a mutualism. *Science* **188**:936–937.

Janzen, D. H. 1975b. Intra- and interhabitat variations in *Guazuma ulmifolia* (Sterculiaceae) seed predation by *Amblycerus cistalinus* (Bruchidae) in Costa Rica. *Ecology* **56**:1009–1013.

Janzen, D. H. 1976a. Why bamboos wait so long to flower. *Annu. Rev. Ecol. Syst.* **7**:347–391.

Janzen, D. H. 1976b. Why tropical trees have rotten cores. *Biotropica* **8**:110.

Janzen, D. H. 1977a. A note on optimal mate selection by plants. *Am. Nat.* **111**:365–374.

Janzen, D. H. 1977b. Why fruits rot, seeds mold, and meat spoils. *Am. Nat.* **111**:691–713.

Janzen, D. H. 1979a. How many babies do figs pay for babies? *Biotropica* **11**:48–50.

Janzen, D. H. 1979b. How to be a fig. *Annu. Rev. Ecol. Syst.* **10**:13–51.

Janzen, D. H. 1980a. When is it coevolution? *Evolution* **34**:611–612.

Janzen, D. H. 1980b. Specificity of seed-attacking beetles in a Costa Rican deciduous forest. *J. Ecol.* **68**:929–952.

Janzen, D. H. 1981. The peak in North American ichneumonid species richness lies between 38° and 42° N. *Ecology* **62**:532–537.

Janzen, D. H., and C. M. Pond. 1975. A comparison, by sweep sampling, of the arthopod fauna of secondary vegetation in Michigan, England, and Costa Rica. *Trans. R. Ent. Soc. Lond.* **127**:33–50.

Janzen, D. H., G. A. Miller, J. Hackforth-Jones, C. M. Pond, K. Hooper, and D. P. Janos. 1976. Two Costa Rican bat-generated seed shadows of *Andira inermis* (Leguminosae). *Ecology* **57**:1068–1075.

Jeffords, M. R., J. G. Sternberg, ad G. P. Waldbauer. 1979. Batesian mimicry: field demonstration of the survival value of pipevine swallotail and monarch color patterns. *Ecology* **33**:275–286.

Jennings, J. B. 1974. Symbioses in the Turbellaria and their implications in studies on the evolution of parasitism. Pages 127–160 in W. B. Vernberg, Ed. *Symbiosis in the sea.* University of South Carolina Press, Columbia.

Johnson, L. K., and S. P. Hubbell. 1975. Contrasting foraging strategies and coexistence of two bee species on a single resource. *Ecology* **56**:1398–1406.

Jones, J. S., B. H. Leith, and P. Rawlings. 1977. Polymorphism in *Cepaea:* a problem with too many solutions? *Annu. Rev. Ecol. Syst.* **8**:109–143.

Karplus, I. 1981. Goby-shrimp partner specificity. II. The behavioural mechanisms regulating partner specificity. *J. Exp. Mar. Biol. Ecol.* **51**:21–35.

Karplus, I., R. Szlep, and M. Tsurnamal. 1981. Goby-shrimp partner specificity. I. Distribution in the northern Red Sea and partner specificity. *J. Exp. Mar. Biol. Ecol.* **51**:1–19.

Karr, J. R. 1980. Turnover dynamics in a tropical continental avifauna. *Proc. 17th Int. Ornit. Congr.* pp. 764–769.

Kaufman, T. 1965. Biological studies of some Bavarian Acridoidea (Orthoptera) with species reference to their feeding habits. *Ann. Entomol. Soc. Am.* **58**:791–800.

Kevan, P. G., W. G. Chaloner, and D. B. O. Saville. 1975. Interrelationships of early terrestrial arthropods and plants. *Palaeontology* **18**:391–417.

Kitting, C. L. 1980. Herbivore–plant interactions of individual limpets maintaining a mixed diet of intertidal marine algae. *Ecol. Monogr.* **50**:527–50.

Kleinfeldt, S. E. 1978. Ant-gardens: the interactions of *Codonanthe crassifolia* (Gesneriaceae) and *Crematogaster longispina* (Formicidae). *Ecology* **59**:449–456.

Kloet, G. S., and W. D. Hincks. 1964. *A check list of British insects; second edition. Part 1.* Royal Entomological Entomological Society, London.

Knerer, G., and C. Atwood. 1973. Diprionid sawflies: polymorphism and speciation. *Science* **179**:1090–1099.

Kohn, A. J. 1978. Ecological shift and release in an isolated population: *Conus miliaris* at Easter Island. *Ecol. Monogr.* **48**:323–336.

Lacey, J. C., Jr., A. L. Weber, and W. E. White, Jr. 1975. A model for the co-evolution of the genetic code and the process of protein synthesis: review and assessment. *Orig. Life* **6**:273–283.

Lack, D. 1965. Evolutionary ecology. *J. Anim. Ecol.* **34**:223–231.

Larson, R. J. 1980. Competition, habitat selection and the bathymetric segregation of two rockfish *(Sebastes)* species. *Ecol. Monogr.* **50**:221–239.

Lawlor, L. R. 1980. Structure and stability in natural and randomly constructed competitive communities. *Am. Nat.* **116**:394–408.

Lawlor, L. R., and J. Maynard-Smith. 1976. The coevolution and stability of competing species. *Am. Nat.* **110**:79–99.

Lawton, J. H. 1978. Host–plant influences on insect diversity: the effects of space and time. Pages 105–125 in L.A. Mound, and N. Waloff, Eds. *Diversity of insect faunas. Symp. R. Ent. Soc. Lond. No. 9.* Blackwell, Oxford.

Lawton, J. H., and M. P. Hassell. 1981. Asymmetrical competition in insects. *Nature* **289**:793–795.

Lawton, J. H., and D. R. Strong, Jr. 1981. Community patterns and competition in folivorous insects. *Am. Nat.* **118**:317–338.

Leius, J. 1960. Attractiveness of different foods and flowers to the adults of some hymenopterous parasites. *Can. Entomol.* **92**:369–376.

Leius, K. 1961a. Influence of food on fecundity and longevity of adults of *Itoplectis conquisitor* (Say) (Hymenoptera: Ichneumonidae). *Can. Entomol.* **93**:771–780.

Leius, K. 1961b. Influence of various foods on fecundity and longevity of adults of *Scambus buolianae* (Htg.) (Hymenoptera: Ichneumonidae). *Can. Entomol.* **93**:1079–1084.

Leius, K. 1963. Effects of pollens on fecundity and longevity of adult *Scambus buolianae* (Htg.) (Hymenoptera: Ichneumonidae). *Can. Entomol.* **95**:202–207.

Leius, K. 1967. Food sources and preferences of adults of a parasite, *Scambus buolinae* (Hym.: Ichn.), and their consequences. *Can. Entomol.* **99**:865–871.

Leppik, E. E. 1975. Morphogenic stagnation in the evolution of *Magnolia* flowers. *Phytomorphology* **25**:451–464.

Levin, D. A. 1979. Pollinator foraging behavior: genetic implications for plants. Pages 131–153 in O. T. Solbrig, S. Jain, G. B. Johnson, and P. H. Raven, Eds. *Topics in plant population biology.* Columbia University Press, New York.

Levin, D. A., and H. W. Kerster. 1974. Gene flow in seed plants. *Evol. Biol.* **7**:139–220.

Levin, S. A., and R. T. Paine. 1974. Disturbance, patch formation and community structure. *Proc. Natl. Acad. Sci. USA* **71**, pp. 2744–2747.

Levin, S. A., and J. D. Udovic. 1977. A mathematical model of coevolving populations. *Am. Nat.* **111**:657–675

Levine, S. H. 1976. Competitive interactions in ecosystems. *Am. Nat.* **110**:903–910.

Levins, R., and R. H. MacArthur. 1969. An hypothesis to explain the incidence of monophagy. *Ecology* **50**:910–911.

Lewis, D. H. 1973. The relevance of symbiosis to taxonomy and ecology, with particular reference to mutualistic symbioses and the exploitation of marginal habitats. Pages 151–172 in V. H. Heywood, Ed. *Taxonomy and ecology.* Academic, New York.

Lewontin, R. C. 1974. *The genetic basis of evolutionary change.* Columbia University Press, New York.

Ligon, J. D. 1978. Reproductive interdependence of Pinon Jays and pinon pines. *Ecol. Monogr.* **48**:111–126.

Limbaugh, C. 1961. Cleaning symbiosis. *Sci. Am.* **205**(2):42–49.

Lin, N., and C. D. Michener. 1972. Evolution of sociality in insects. *Quart. Rev. Biol.* **47**:131–159.

Linhart, Y. B., and P. Feinsinger. 1980. Plant–hummingbird interactions: effects of island size and degree of specialization on pollination. *J. Ecol.* **68**:745–760.

Linsley, E. G. 1958. The ecology of solitary bees. *Hilgardia.* **27**:543–599.

Linsley, E. G., and J. W. MacSwain. 1957. The nesting habits, flower relationships, and parasites of some North American species of *Diadasia*. *Wasmann J. Biol.* **15**:199–235.

Linsley, E. G., J. W. MacSwain, P. H. Raven, and R. W. Thorp. 1973. Comparative behavior of bees and Onagraceae. V. *Camissonia* and *Oenothera* bees of cismontane California and Baja California. *Univ. Calif. Pub. Entomol.* **71**:1–68.

Lister, B. C. 1980. Resource variation and the structure of British bird communities. *Proc. Natl. Acad. Sci., USA* **77**, pp. 4185–4187.

Lloyd, M., and H. S. Dybas. 1966. The periodical cicada problem. I. Population ecology. *Evolution* **20**:133–149.

Lobel, P., and J. C. Ogden. 1981. Foraging by the herbivorous parrotfish *Sparisoma radians*. *Mar. Biol.* **64**:173–183.

Lomnicki, A. 1971. Animal population regulation by the genetic feedback mechanism: a critique of the theoretical model. *Am. Nat.* **105**:413–421.

Losey, G. S. Jr. 1971. Communication between fishes in cleaning symbiosis. Pages 45–76 in T. C. Cheng, Ed. *Aspects of the biology of symbiosis.* University Park Press, Baltimore.

Losey, G. S., Jr. 1979. Fish cleaning symbiosis: proximate causes of host behaviour. *Anim. Behav.* **27**:669–685.

Lowe, E. F., and J. M. Lawrence. 1976. Absorption efficiencies of *Lytechinus variegatus* (Lamarck) (Echinodermata: Echinoidea) for selected marine plants. *J. Exp. Mar. Biol. Ecol.* **21**:223–234.

Lubchenko, J. 1979. Consumer terms and concepts. *Am. Nat.* **113**:315–317.

Lubchenko, J. 1980. Algal zonation in the New England rocky intertidal community: an experimental analysis. *Ecology* **61**:333–334.

MacArthur, R. H. 1958. Population ecology of some warblers of northeastern coniferous forests. *Ecology* **39**:599–619.

MacArthur, R. H. 1972. *Biographical ecology: patterns in the distribution of species.* Harper & Row, New York.

Macedo, M., and G. T. Prance. 1978. Notes on the vegetation of Amazonia II. The dispersal of plants in Amazonian white sand campinas: the campinas as functional islands. *Brittonia* **30**:203–215.

Mack, R. N., and J. N. Thompson. 1982. Evolution in steppe with few large hooved mammals. *Am. Nat.* **119**:757–773.

MacSwain, J. W., P. H. Raven, and R. W. Thorp. 1973. Comparative behavior of bees and Onagraceae. IV. *Clarkia* bees of the western United States. *Univ. Calif. Publ. Entomol.* **70:**1–80.

Maiorana, V. C. 1977. Tail autotomy, functional conflicts and their resolution by a salamander. *Nature* **265:**533–535.

Maiorana, V. C. 1978. Nontoxic toxins: the energetics of coevolution. *Biol. J. Linn. Soc.* **11:**387–396.

Major, P. F. 1973. Scale feeding behavior of the leatherjacket *Scombroides layson,* and two species of the genus *Oligoplites.* (Pisces:Carangidae). *Copeia* **1973:** 151–154.

Maser, C., J. M. Trappe, and R. A. Nussbaum. 1978. Fungal–small mammal interrelationships on Oregon coniferous forests. *Ecology* **59:**799–809.

May, R. M. 1973. *Stability and complexity in model ecosystems.* Princeton University Press, Princeton, NJ.

May, R. M. 1976. Models for two interaction populations. Pages 49–70 in R. M. May, Ed. *Theoretical ecology: principles and applications.* Saunders, Philadelphia.

May, R. M., Ed. 1981. *Theoretical ecology: principles and applications,* Second edition. Saunders, Philadelphia.

Mayr, E. 1970. *Populations, species, and evolution.* Harvard University Press, Cambridge, MA.

McBee, R. H. 1971. Significance of intestinal microflora in herbivory. *Annu. Rev. Ecol. Syst.* **2:**165–176.

McClure, M. S. 1977. Population dynamics of the red pine scale *Matsucoccus resinosae:* the influence of resinosis. *Environ. Entomol.* **6:**789–795.

McClure, M. S., and P. W. Price. 1975. Competition among sympatric *Erythroneura* leafhoppers (Homoptera: Cicadellidae) on American sycamore. *Ecology* **56:**1388–1397.

McDade, L. A., and S. Kinsman. 1980. The impact of floral parasitism in two neotropical hummingbird-pollinated plant species. *Evolution* **34:**944–958.

McKey, D. 1974. Adaptive patterns in alkaloid physiology. *Am. Nat.* **108:**305–320.

McKey, D. 1975. The ecology of coevolved seed dispersal systems. Pages 159–191 in L. E. Gilbert and P. H. Raven, Eds. *Coevolution of animals and plants.* University of Texas Press, Austin.

McKey, D. 1979. The distribution of secondary compounds within plants. Pages 55–133 in G. A. Rosenthal and D. H. Janzen, Eds. *Herbivores: their interaction with secondary plant metabolites.* Academic, New York.

McNaughton, S. J. 1978. Serengeti ungulates: feeding selectivity influences the effectiveness of plant defense guilds. *Science* **199:**806–807.

McNaughton, S. J. 1979. Grazing as an optimization process: grass–ungulate relationships in the Serengeti. *Am. Nat.* **113:**691–703.

Medawar, P. B. 1957. *The uniqueness of the individual.* Basic, New York.

Menge, B. A. 1972. Competition for food between two intertidal starfish species and its effect on body size and feeding. *Ecology* **53:**635–644.

Menge, B. A. 1979. Coexistence between the seastars *Asterias vulgaris* and *A. forbesi* in a heterogeneous environment: a non-equilibrium explanation. *Oecologia* **41:**245–272.

Menge, B. A., and J. P. Sutherland. 1976. Species diversity gradients: synthesis of the roles of predation, competition and temporal heterogeneity. *Am. Nat.* **110:**351–369.

Merchant, H. 1972. Estimated population size and home range of the salamanders *Plethodon jordani* and *Plethodon glutionosus. J. Wash. Acad. Sci.* **62:**248–257.

Miller, R. S. 1967. Pattern and process in competition. *Adv. Ecol. Res.* **4**:1–74.

Milton, K. 1980. *The foraging strategy of Howler Monkeys: a study of primate economics.* Columbia University Press, New York.

Mode, C. J. 1958. A mathematical model for the co-evolution of obligate parasites and their hosts. *Evolution* **12**:158–165.

Montgomery, G. G., Ed. 1978. *The ecology of arboreal folivores.* Smithsonian Institution Press, Washington, DC.

Montgomery, G. G., and M. E. Sunquist. 1975. Impact of sloths on neotropical forest energy flow and nutrient cycling. Pages 69–98 in F. B. Golley and E. Medina, Eds. *Tropical ecological systems: trends in terrestrial and aquatic research.* Springer-Verlag, New York.

Montgomery, G. G., and M. E. Sunquist. 1978. Habitat selection and use by two-toed and three-toed sloths. Pages 329–359 in G. G. Montgomery, ed. *The ecology of arboreal folivores.* Smithsonian Institution Press, Washington, DC.

Moore, H. E. 1973. A synopsis of the genus *Codonanthe* (Gesneriaceae). *Baileya* **19**:4–33.

Morrow, P. A., and L. R. Fox. 1980. Effects of variation in *Eucalyptus* essential oil yield on insect growth and grazing damage. *Oecologia* **45**:209–219.

Morse, D. H. 1971. The insectivorous bird as an adaptive strategy. *Annu. Rev. Ecol. Syst.* **2**:177–200.

Morton, E. S. 1973. On the evolutionary advantages and disadvantages of fruit eating in tropical birds. *Am. Nat.* **107**:8–22.

Mulcahy, D. L. 1979. The rise of angiosperms: a genecological factor. *Science* **206**:20–23.

Munger, J. C., and J. H. Brown. 1981. Competition in desert rodents: an experiment with semipermeable exclosures. *Science* **211**:510–12.

Murdoch, W. W., and A. Oaten. 1975. Predation and population stability. *Adv. Ecol. Res.* **9**:1–131.

Nault, L. R., and D. M. DeLong. 1980. Evidence for coevolution of leafhoppers in the genus *Dalbulus* (Cicadellidae: Homoptera) with maize and its ancestors. *Ann. Ent. Soc. Am.* **73**:349–353.

Noirot, C., and C. Noirot-Timothea. 1969. The digestive system. Pages 49–88 in K. Khrisna and F. M. Weesner, Eds. *Biology of termites.* Volume 1. Academic, New York.

Orians, G. H. 1962. Natural selection and ecological theory. *Am. Nat.* **96**:257–263.

Owen, D. F. 1980. How plants may benefit from the animals that eat them. *Oikos* **35**:230–235.

Owen, D. F., and J. Owen. 1974. Species diversity in temperate and tropical Ichneumonidae. *Nature* **240**:583–584.

Owen, D. F., and R. G. Wiegert. 1976. Do consumers maximize plant fitness? *Oikos* **27**:488–492.

Owen, D. F., and R. G. Wiegert. 1981. Mutualism between grasses and grazers: an evolutionary hypothesis. *Oikos* **36**:376–378.

Paine, R. T. 1966. Food web complexity and species diversity. *Am. Nat.* **100**:65–75.

Paine, R. T. 1969. The *Pisaster–Tegula* interaction: prey patches, predator food preference and interidal community structure. *Ecology* **50**:950–961.

Paine, R. T. 1974. Intertidal community structure: experimental studies on the relationship between a dominant competitor and its principal predator. *Oecologia* **15**:93–120.

Paine, R. T. 1977. *Controlled manipulation in the marine intertidal zone and their contributions to ecological theory*. Academy of Natural Sciences, Philadelphia, Special Publication No. 12, pp. 245–270.

Paine, R. T. 1980. Food webs: linkage, interaction strength and community infrastructure. *J. Anim. Ecol.* **49**:667–685.

Painter, R. H. 1951. *Insect resistance in crop plants*. Macmillan, New York.

Pearson, D. L., and E. J. Mury. 1979. Character divergence and convergence among tiger beetles (Coleoptera: Cicindellidae). *Ecology* **60**:557–566.

Person, C. O. 1967. Genetic aspects of parasitism. *Can. J. Bot.* **45**:1193–1204.

Person, C., D. J. Samborski, and R. Rohringer. 1962. The gene-for-gene concept. *Nature* **199**:561–562.

Petelle, M. 1980. Aphids and melezitose: a test of Owen's 1978 hypothesis. *Oikos* **35**:127–128.

Pianka, E. R., R. B. Huey, and L. R. Lawlor. 1979. Niche segregation in desert lizards. Pages 67–115 in D. J. Horn, G. R. Stairs, and R. D. Mitchell Eds. *Analysis of ecological systems*. Ohio State University Press, Columbus.

Pickett, S. T. A. 1976. Succession: an evolutionary perspective. *Am. Nat.* **110**:107–119.

Pickett, S. T. A. 1980. Non-equilibrium coexistence of plants. *Bull. Torrey Bot. Club* **107**:238–248.

Pickett, S. T. A., and J. N. Thompson. 1978. Patch dynamics and the design of nature reserves. *Biol. Conserv.* **13**:27–37.

Pielou, E. C. 1969. *An introduction to mathematical ecology*. Wiley, New York.

Pierce, N. E., and P. S. Mead. 1981. Parasitoids as selective agents in the symbiosis between lycaenid butterfly larvae and ants. *Science* **211**:1185–1187.

Pimentel, D. 1961. Animal population regulation by the genetic-feedback mechanism. *Am. Nat.* **95**:65–79.

Pimentel, D., E. H. Feinburg, D. W. Wood, and J. T. Hayes. 1965. Selection, spatial distribution and the coexistence of competing fly species. *Am. Nat.* **99**:97–108.

Pimentel, D., S. A. Levin, and D. Olson. 1978. Coevolution and the stability of exploiter–victim systems. *Am. Nat.* **112**:119–125.

Platt, W. J., G. R. Hill, and S. Clark. 1974. Seed predation in a prairie legume (*Astragalus canadensis* L.): interactions between pollination, predispersal seed predation, and plant density. *Oecologia* **17**:55–63.

Powell, J. A. 1980. Evolution of larval food preferences in Microlepidoptera. *Annu. Rev. Entomol.* **25**:133–159.

Powell, J. A., and R. A. Mackie. 1966. Biological interrelationships of moths and *Yucca whipplei* (Lepidoptera:Gelechiidae, Blastobasidae, Prodoxidae). *Univ. Calif. Publ. Entomol.* **42**:1–59.

Prance, G. T. 1973. Gesneriads in the ant gardens of the Amazon. *Gloxinian* **23**:27–28.

Price, P. W. 1972. Parasitoids utilizing the same host: adaptive nature of differences in size and form. *Ecology* **53**:129–140.

Price, P. W. 1975a. *Insect ecology*. Wiley–Interscience, New York.

Price, P. W. 1975b. Introduction: the parasitic way of life and its consequences. Pages 1–13 in P. W. Price, Ed. *Evolutionary strategies of parasitic insects and mites*. Plenum, New York.

Price, P. W. 1977. General concepts in the evolutionary biology of parasites. *Evolution* **31**:405–420.

156 REFERENCES

Price, P. W. 1980. *Evolutionary biology of parasites*. Princeton University Press, Princeton, NJ.

Price, P. W., C. E. Bouton, P. Gross, B. A. McPheron, J. N. Thompson, and A. E. Weis. 1980. Interactions among three trophic levels: influence of plants on interactions between insect herbivores and natural enemies. *Annu. Rev. Ecol. Syst.* **11**:41–65.

Pudlo, R. J., A. J. Beattie, and D. C. Culver. 1980. Population consequences of changes in an ant–seed mutualism in *Sanguinaria canadensis. Oecologia* **146**:32–37.

Pulliam, H. R. 1975. Diet optimization with nutrient constraints. *Am. Nat.* **109**:765–768.

Pyke, G. H., H. R. Pulliam, and E. L. Charnov. 1977. Optimal foraging: a selective review of theory and tests. *Quart. Rev. Biol.* **52**:137–154.

Ralph, C. P. 1977. Effect of host plant density on populations of a specialized, seed-sucking bug, *Oncopeltus fasciatus. Ecology* **58**:799–809.

Ramírez, W. 1970. Host specificity of fig wasps *(Agaonidae). Evolution* **24**:680–691.

Ramírez, W. 1974. Coevolution of *Ficus* and *Agaonidae. Ann. Missouri Bot. Gard.* **61**:770–780.

Ramírez, B. W. 1976. Evolution of blastophagy. *Brenesia* **9**:1–14.

Rapport, D. J. 1980. Optimal foraging for complementary resources. *Am. Nat.* **116**:324–346.

Rathcke, B. J., and P. W. Price. 1976. Anomalous diversity of tropical ichneumonid parasitoids: a predation hypothesis. *Am. Nat.* **110**:889–893.

Rausher, M. D. 1979. Larval habitat suitability and oviposition preference in three related butterflies. *Ecology* **60**:503–511.

Rausher, M. D. 1980. Host abundance, juvenile survival, and oviposition preference in *Battus philenor. Evolution.* **34**:342–355.

Raven, P. H. 1977. A suggestion concerning the Cretaceous rise to dominance of the angiosperms. *Evolution* **31**:451–452.

Regal, P. J. 1977. Ecology and evolution of flowering plant dominance. *Science* **196**:622–629.

Reynoldson, T. B., and L. S. Bellamy. 1971. The establishment of interspecific competition in field populations, with an example of competition in action between *Polycelis nigra* (Mull.) and *P. tenuis* (Ijima) (Turbellaria, Tricladida). Pages 282–297 in P. J. den Boer and G. R. Gradwell, Eds. *Dynamics of populations*. Centre for Agricultural Publishing and Documentation, Wageningen, the Ntherlands.

Rhoades, D. F. 1979. Evolution of plant chemical defenses against herbivores. Pages 3–54 in G. A. Rosenthal and D. H. Janzen, Eds. *Herbivores: their interaction with secondary plant metabolites*. Academic, New York.

Rhoades, D. F., and R. G. Cates. 1976. Toward a general throry of plant antiherbivore chemistry. *Rec. Adv. Phytochem.* **10**:168–213.

Ricklefs, R. E. 1969. An analysis of nesting mortality in birds. *Smithsonian Contrib. Zool.* **9**:1–48.

Ricklefs, R. E. 1974. Energetics of reproduction in birds. Pages 152–202 in R. E. Paynter, Jr., Ed. *Avian Energetics. Publ. Nuttall Ornith. Club*, No. 15.

Ricklefs, R. E. 1976. Growth rates of birds in the humid New World tropics. *Ibis* **118**:179–207.

Rickson, F. R. 1979. Absorption of animal tissue breakdown products into a plant stem—the feeding of a plant by ants. *Am. J. Bot.* **66**:87–90.

Riley, C. V. 1892. The yucca moth and *Yucca* pollination. *Ann. Rep. Missouri Bot. Garden* **3**:99–158.

Roberts, R. C. 1979. The evolution of avian food-storing behavior. *Am. Nat.* **114**:418–438.

Rockwood, L. L. 1976. Plant selection and foraging patterns in two species of leaf-cutting ants *(Atta)*. *Ecology* **57**:48–61.

Root, R. B. 1967. The niche exploitation pattern of the Blue-gray Gnatcatcher. *Ecol. Monogr.* **37**:317–350.

Root, R. B. 1973. Organization of a plant-arthropod association in simple and diverse habitats: the fauna of collards *(Brassica oleracea)*. *Ecol. Monogr.* **43**:95–125.

Ross, G. N. 1966. Life-history studies on Mexican butterflies. IV. The ecology and ethology of *Anatole rossi,* a myrmecophilous metalmark (Lepidoptera:Riodinidae). *Ann. Entomol. Soc. Am.* **59**:985–1004.

Roth, V. L. 1981. Constancy in the size ratios of sympatric species. *Am. Nat.* **118**:394–404.

Rothschild, M., and T. Clay. 1952. *Fleas, flukes and cuckoos. A study of bird parasites.* Collins, London.

Roughgarden, J. 1975. Evolution of marine symbiosis: a simple cost-benefit model. *Ecology* **56**:1201–1208.

Roughgarden, J. 1976. Resource partitioning among competing species—a coevolutionary approach. *Theor. Pop. Biol.* **9**:388–424.

Roughgarden, J. 1979. *Theory of population genetics and evolutionary ecology: an introduction.* Macmillan, New York.

Ryan, C. A. 1979. Proteinase inhibitors. Pages 599–618 in G. A. Rosenthal and D. H. Janzen, Eds. *Herbivores: their interactions with secondary plant metabolites.* Academic, New York.

Sale, P. F. 1975. Patterns of use of space in a guild of territorial reef fishes. *Mar. Biol.* **29**:89–97.

Sale, P. R. 1977. Maintenance of high diversity in coral reef fish communities. *Am. Nat.* **111**:337–359.

Sale, P. R. 1978. Chance patterns of demographic change in coral rubble patches at Heron reef. *J. Exp. Mar. Biol. Ecol.* **34**:233–243.

Sale, P. F. 1979. Recruitment, loss and coexistence in a guild of territorial coral reef fishes. *Oecologia* **42**:159–177.

Salt, G. 1970. *The cellular defense reactions of insects.* Cambridge University Press, London.

Schemske, D. W. 1980. The evolutionary significance of extraflora nectar production by *Costus woodsonii* (Zingiberaceae): an experimental analysis of ant protection. *J. Ecol.* **68**:959–967.

Schemske, D. W. 1981. Floral convergence and pollinator sharing in two bee-pollinated tropical herbs. *Ecology* **62**:946–954.

Schemske, D. W. 1982. Ecological correlates of a neotropical mutualism: ant assemblages at *Costus* extrafloral nectaries. *Ecology* **63** (in press).

Schemske, D. W., and N. Brokaw. 1981. Treefalls and the distribution of understory birds in a tropical forest. *Ecology* **62**:938–945.

Schmitt, J. 1980. Pollinator foraging behavior and gene dispersal in *Senecio* (Compositae). *Evolution* **34**:934–943.

Schoener, T. W. 1965. The evolution of bill size differences among sympatric congeneric species of birds. *Evolution* **19**:189–213.

Schoener, T. W. 1971. Theory of feeding strategies. *Annu. Rev. Ecol. Syst.* **2**:369–404.

Schoener, T. W. 1974. Resource partitioning in ecological communities. *Science* **185**:27–39. **185**:27–39.

Schorger, A. W. 1955. *The Passenger Pigeon: its natural history and extinction.* University of Wisconsin Press, Madison.

Schwartz, P., and D. W. Snow. 1978. Display and related behavior of the Wire-tailed Manakin. *Living Bird* **17**:51–77.

Scriber, J. M. 1973. Latitudinal gradients in larval feeding specialization of the world Papilionidae (Lepidoptera). *Psyche* **80**:355–373.

Scriber, J. M. 1977. Limiting effects of low leaf-water content on the nitrogen utilization, energy budget, and larval growth of *Hyalophora cecropia* (Lepidoptera: Saturniidae). *Oecologia* **28**:269–287.

Scriber, J. M. 1978. The effects of larval feeding specialization and plant growth form on the consumption and utilization of plant biomass and nitrogen: an ecological consideration. *Entomol. Exp. Appl.* **24**:494–510.

Scriber, J. M. 1979. Post-ingestive utilization of plant biomass and nitrogen by Lepidoptera: legume feeding by the southern armyworm. *N.Y. Entomol. Soc.* **87**:141–153.

Scriber, J. M., and F. Slansky, Jr. 1981. The nutritional ecology of immature insects. *Annu. Rev. Entomol.* **26**:183–211.

Sick, H. 1967. Courtship behavior in the manakins (Pipridae): a review. *Living Bird* **6**:5–22.

Sidhu, G. S., and J. M. Webster. 1981. The genetics of plant–nematode parasitic systems. *Bot. Rev.* **47**:387–419.

Silvertown, J. W. 1980. The evolutionary ecology of mast seeding in trees. *Biol. J. Linn. Soc.* **14**:235–250.

Simberloff, D. S. 1976. Trophic structure determination and equilibrium in an arthropod community. *Ecology* **57**:395–398.

Simberloff, D., B. J. Brown, and S. Lowrie. 1978. Isopod and insect root borers may benefit Florida mangroves. *Science* **201**:630–632.

Simberloff, D., and E. F. Connor. 1981. Missing species combinations. *Am. Nat.* **118**:215–239.

Singer, M. C. 1971. Evolution of food-plant preference in the butterfly *Euphydryas editha.* *Evolution* **25**:383–389.

Singer, M. C. 1972. Complex components of habitat suitability within a butterfly colony. *Science* **176**:75–77.

Skinner, G. J., and J. B. Whittaker. 1981. An experimental investigation of inter-relationships between the wood-ant *(Formica rufa)* and some tree-canopy heribivores. *J. Anim. Ecol.* **50**:313–326.

Skutch, A. F. 1949. Life history of the Yellow-Thighed Manakin. *Auk* **66**:1–24.

Skutch, A. F. 1960. Life histories of Central American birds. II. *Pac. Coast Avif.*, No. 34.

Skutch, A. F. 1980. *A naturalist on a tropial farm.* University of California Press, Berkeley.

Slansky, F., Jr. 1976. Phagism relationships among butterflies. *J. N.Y. Entomol. Soc.* **84**:91–105.

Slatkin, M. 1980. Ecological character displacement. *Ecology* **61**:163–177.

Slatkin, M., and J. Maynard Smith. 1979. Models of coevolution. *Quart. Rev. Biol.* **54**:233–263.

Slatkin, M., and D. S. Wilson. 1979. Coevolution in structured demes. *Proc. Natl. Acad. Sci. USA* **76**, pp. 2084–2087.

Smiley, J. 1978. Plant chemistry and evolution of host specificity: new evidence from *Heliconius* and Passiflora. *Science* 201:745–747.

Smith, C. C. 1970. The coevolution of pine squirrels *(Tamiasciurus)* and conifers. *Ecol. Monogr.* 40:349–371.

Smith, C. C. 1975. The coevolution of plants and seed predators. Pages 53–77 in L. E. Gilbert and P. H. Raven, Eds. *Coevolution of Animals and Plants.* University of Texas Press, Austin.

Smith, C. C., and R. P. Balda. 1979. Competition among insects, birds and mammals for conifer seeds. *Am. Zool.* 19:1065–1083.

Smith, D. C. 1981. Competitive interactions of the striped plateau lizard *(Sceloporus virgatus)* and the tree lizard *(Urasaurus ornatus).* *Ecology* 62:679–687.

Smith, N. G. 1968. The advantage of being parasitized. *Nature* 219:690–694.

Smith, N. G. 1979. Alternate responses by hosts to parasites which may be helpful or harmful. Pages 7–15 in B. B. Nickol. (Ed.). *Host–parasite interfaces.* Academic, New York.

Smith, N. G. 1980. Some evolutionary, ecological, and behavioural correlates of communal meeting by birds with wasps or bees. *Proc. 17th Int. Ornith. Congr.,* pp. 1199–1205.

Smith, S. M. 1977. Coral-snake pattern recognition and stimulus generalization by naive great kiskadees (Aves: Tyrannidae). *Nature* 265:535–536.

Smythe, N. 1970. Relationships between fruiting seasons and seed dispersed methods in a Neotropical forest. *Am. Nat.* 104:25–35.

Snow, B. K. 1970. A field study of the Bearded Bellbird in Trinidad. *Ibis* 112:299–329.

Snow, B. K., and D. W. Snow. 1971. The feeding ecology of Tanagers and honeycreepers in Trinidad. *Auk* 88:291–322.

Snow, B. K., and D. W. Snow. 1979. The Ochre-bellied Flycatcher and the evolution of lek behavior. *Condor* 81:286–292.

Snow, D. W. 1962a. A field study of the Black and White Manakin, *Manacus manacus,* in Trinidad. *Zoologica* 47:65–104.

Snow, D. W. 1962b. A field study of the Golden-headed Manakin, *Pipra erythrocephala,* in Trinidad, W. I. *Zoologica* 47:183–198.

Snow, D. W. 1962c. The natural history of the Oilbird, *Steatornis caripensis,* in Trinidad, W. I. Part. 2. Population, breeding ecology and food. *Zoologica* 47:199–221.

Snow, D. W. 1971. Evolutionary aspects of fruit-eating by birds. *Ibis* 113:194–202.

Snow, D. W. 1980. Regional differences between tropical floras and the evolution of frugivory. *Proc. 17th Int. Congr. Ornith.,* pp. 1192–1198.

Snow, D. W. 1981. Tropical frugivorous birds and their food plants: a world survey. *Biotropica* 13:1–14.

Snyder, N. F. R., and H. A. Snyder. 1969. A comparative study of mollusc predation by Limpkins, Everglade Kites, and Boat-Tailed Grackles. *Liv. Bird* 8:177–223.

Sorensen, A. E. 1981. Interactions between birds and fruit in a temperate woodland. *Oecologia* 50:242–249.

Sorenson, D., and W. T. Jackson, 1968. Utilization of paramecium by *Utricularia gibba.* *Planta* 83:166–170.

Springett, B. P. 1968. Aspects of the relationship between burying beetles, *Necrophorus* spp. and the mite *Poecilochirus necrophori* Vitz. *J. Anim. Ecol.* 37:417–424.

Stahl, E. 1884. Pflanzen und Schnecken. *Jena Z. Med. Naturwiss.* 22:557–584.

Stapanian, M. A., and C. C. Smith. 1978. A model for seed scatterhoarding: coevolution of fox squirrels and black walnuts. *Ecology* **59**:884–896.

Starmer, W. T., H. W.Kirscher, and H. J. Phaff. 1980. Evolution and speciation of host plant specific yeasts. *Evolution* **34**:137–146.

Starr, M. P. 1975. A generalized scheme for classifying organismic associations. Pages 1–20 in D. H. Jennings, and D. L. Lee. Eds. *Symbiosis. Symp. Soc. Exp. Biol. No. 29.* Cambridge University Press, Cambridge.

Stebbins, G. L. 1971. *Chromosomal evolution in higher plants.* Arnold, London.

Stebbins, G. L. 1981. Coevolution of grasses and herbivores. *Ann. Missouri Bot. Gard.* **68**:75–86.

Stenseth, N. C. 1978. Do grazers maximize individual plant fitness? *Oikos* **31**:299–306.

Stenseth, N. C. and L. Hansson. 1979. Optimal food selection: a graphical model. *Am. Nat.* **113**:373–389.

Stiles, E. W. 1980. Patterns of fruit presentation and seed dispersal in bird-disseminated woody plants in the eastern deciduous forest. *Am. Nat.* **116**:670–688.

Stiling, P. D. 1980. Competition and coexistence among *Eupteryx* leafhoppers (Hemiptera: Cicadellidae) occurring on stinging nettles *(Urtica dioica). J. Anim. Ecol.* **49**:793–805.

Strong, D. R., Jr. 1974. The insects of British trees: community equilibrium in ecological time. *Ann. Missouri Bot. Gard.* **61**:692–701.

Strong, D. R., Jr. 1979. Biogeographic dynamics of insect–host plant communities. *Annu. Rev. Entomol.* **24**:89–119.

Strong, D. R., Jr. and D. S. Simberloff. 1981. Straining at gnats and swallowing ratios: character displacement. *Evolution* **35**:810–812.

Strong, D. R., Jr. L. A. Szyska, and D. S. Simberloff. 1979. Tests of community-wide character displacement against null hypotheses. *Evolution* **33**:987–913.

Svardson, G. 1957. The "invasion" type of bird migration. *Brit. Birds* **50**:314–343.

Sykes, P. W., and H. W. Kale II. 1974. Everglade Kites feed on nonsnail prey. *Auk* **91**:818–820.

Tabashnik, B. E., H. Whellock, J. D. Rainbolt, and W. B. Ward. 1981. Individual variation in oviposition preference in the butterfly, *Colias eurytheme. Oecologia* **50**:225–230.

Talbot, F. H., B. C. Russell, and G. R. V. Anderson. 1978. Coral reef fish communities: unstable, high-diversity systems? *Ecol. Monogr.* **48**:425–440.

Talbot, M. 1934. Distribution of ant species in the Chicago region with reference to ecological factors and physiological tolerances. *Ecology* **15**:416–439.

Taylor, R. J. 1974. Role of learning in insect parasitoids. *Ecol. Monogr.* **44**:89–104.

Terborgh, J. 1977. Bird species diversity on an Andean elevational gradient. *Ecology* **58**:1007–1019.

Terborgh, J. and J. Faaborg. 1980. Saturation of bird communities in the West Indies. *Am. Nat.* **116**:178–195.

Thompson, J. N. 1978. Within-patch structure and dynamics in *Pastinaca sativa* and resource availability to a specialized herbivore. *Ecology* **59**:443–448.

Thompson, J. N. 1980. Treefalls and colonization patterns of temperate forest herbs. *Am. Midl. Nat.* **104**:176–184.

Thompson, J. N. 1981a. Elaiosomes and fleshy fruits: phenology and selection pressures for ant-dispersed seeds. *Am. Nat.* **117**:104–108.

Thompson, J. N. 1981b. Reversed animal–plant interactions: the evolution of insectivorous and ant-fed plants. *Biol. J. Linn. Soc.* **16**:147–155.

Thompson, J. N., and P. W. Price. 1977. Plant plasticity, phenology, and herbivore dispersion: wild parsnip and the parsnip webworm. *Ecology* **58**:1112–1119.

Thompson, J. N., and M. F. Willson. 1978. Disturbance and the dispersal of fleshy fruits. *Science* **200**:1161–1163.

Thompson, J. N. and M. F. Willson. 1979. Evolution of temperate fruit/bird interactions: phenological strategies. *Evolution.* **33**:973–982.

Thomson, J. D. 1980. Implications of different sorts of evidence for competition. *Am. Nat.* **116**:719–726.

Thornhill, R. 1980. Competition and coexistence among *Panorpa scorpionflies* (Mecoptera: Panorpidae). *Ecol. Monogr.* **50**:179–197.

Thorp, R. W. 1969. Systematics and ecology of bees of the subgenus *Diandrena*. *Univ. Calif. Publ. Entomol.* **52**:1–146.

Thorsteinson, A. J. 1960. Host selection in phytophagous insects. *Annu. Rev. Entomol.* **5**:193–218.

Tilman, D. 1978. Cherries, ants and tent caterpillars: timing of nectar production in relation to susceptibility of caterpillars to ant predation. *Ecology* **59**:686–692.

Tinnin, R. O. 1972. Interference or competition? *Am. Nat.* **106**:672–675.

Tomback, D. F. 1978. Foraging strategies of Clark's Nutcracker. *Liv. Bird* **16**:123–161.

Trelease, W. 1893. Further studies of yuccas and their pollination. *Ann. Rep. Missouri Bot. Garden.* **4**:181–226.

Turner, D. C. 1975. *Vampire bat: a field study in behavior and ecology.* Johns Hopkins University Press, Baltimore.

Ueckert, D. N., and R. M. Hansen. 1971. Dietary overlap of grasshoppers on sandhill rangeland in northeastern Colorado. *Oecologia* **8**:276–295.

Ule, E. 1901. Ameisengärten in Amazonasgebiet. *Bot. Jahr.* **30**:45–51.

Ule, E. 1906. Ameisenpflanzen. *Bot. Jahr.* **37**:335–352.

van der Meijden, E. 1979. Herbivore exploitation of a fugitive plant species; local survival and extinction of the cinnabar moth and ragwort in a heterogeneous environment. *Oecologia* **42**:307–323.

Vandermeer, J. 1980. Indirect mutualism: variations on a theme by Stephen Levine. *Am. Nat.* **116**:441–448.

Vandermeer, J., and D. Boucher. 1978. Varieties of mutualistic interactions in population models. *J. Theor. Biol.* **74**:549–558.

Vander Wall, S. B., and R. P. Balda. 1977. Coadaptations of the Clark's Nutcracker and the piñon pine for efficient seed harvest and dispersal. *Ecol. Monogr.* **47**:89–111.

Vander Wall, S. B., and R. P. Balda. 1981. Ecology and evolution of food storage behavior in conifer-seed-caching corvids. *Z. Tierpsychol.* **56**:217–242.

Van Valen, L. 1973. A new evolutionary law. *Evol. Theory* **1**:130.

Waage, J. K. 1979. The evolution of insect/vertebrate associations. *Biol. J. Linn. Soc.* **12**:187–224.

Waddington, C. H. 1975. *The evolution of an evolutionist.* Cornell University Press, Ithaca, NY.

Waldbauer, G. P., and J. G. Sternburg. 1976. Saturniid moths as mimics: an alternative interpretation of attempts to demonstrate mimetic advantage in nature. *Evolution* **29**:650−658.

Walsh, B. D. 1864. On phytophagic varieties and phytophagic species. *Proc. Entomol. Soc. Phila.* **3**:403−430.

Waser, N. M. and M. V. Price. 1981. Pollinator choice and stabilizing selection for flower choice is *Delphinium nelsonii. Evolution* **35**:376−390.

Way, M. J. 1963. Mutualism between ants and honeydew-producng Homoptera. *Annu. Rev. Entomoil.* **8**:307−344.

Westoby, M. 1974. An analysis of diet selection by large generalist herbivores. *Am. Nat.* **109**:290−304.

Westoby, M. 1978. What are the biological bases of varied diets? *Am. Nat.* **112**:627−631.

Wheeler, W. M. 1911. Insect parasitism and its peculiarities. *Pop. Sci. Monthly* **79**:431−449.

Wheelwright, N. T. and G. H. Orians. 1982. Seed dispersal by animals: contrasts with pollen dispersal, problems of terminology, and constraints on coevolution. *Am. Nat.* **119**:402−413.

White, M. J. D. 1968. Models of speciation. *Science* **159**:1065−1070.

White, M. J. D. 1978. *Modes of speciation.* Freeman, San Francisco.

White, P. S. 1979. Pattern, process, and natural disturbance in vegetation. *Bot. Rev.* **45**:229−299.

Whitham, T. G., and C. N. Slobodchikoff. 1981. Evolution by individuals, plant−herbivore interactions, and mosaics of genetic variability: the adaptive significance of somatic mutations in plants. *Oecologia* **49**:287−292.

Wiebes, J. T. 1979. Co-evolution of figs and their pollinators. *Annu. Rev. Ecol. Syst.* **10**:1−12.

Wiens, J. A. 1977. On competition and variable environments. *Am. Sci.* **65**:590−597.

Wiens, J. A., and J. T. Rotenberry. 1981. Morphological size ratios and competition in ecological communities. *Am. Nat.* **117**:592−599.

Wiklund, C. 1981. Generalist vs. specialist oviposition behaviour in *Papilio machaon* (Lepidoptera) and functional aspects on the hierarchy of oviposition preferences. *Oikos* **36**:163−170.

Williams, A. H. 1981. An analysis of competitive interactions in a patchy back-reef environment. *Ecology* **62**:1107−1120.

Williams, G. C. 1957. Pleiotropy, natural selection, and the evolution of senescence. *Evolution* **11**:398−411.

Williams, J. B., and G. O. Batzli. 1979. Competition among bark-foraging birds in Central Illinois: experimental evidence. *Condor* **81**:122−132.

Williams, K. S., and L. E. Gilbert. 1981. Insects as selective agents on plant vegetative morphology: egg mimicry reduces egg laying in butterflies. *Science* **212**:467−469.

Willson, M. F. 1979. Sexual selection in plants. *Am. Nat.* **113**:777−790.

Willson, M. F. and R. I. Bertin. 1979. Flower visitors, nectar production, and inflorescence size of *Asclepias syriaca* L. *Can. J. Bot.* **57**:1380−1388.

Willson, M. F., and B. J. Rathcke. 1974. Adaptive design of the floral display in *Asclepias syriaca* L. *Am. Midl. Nat.* **92**:47−57.

Willson, M. F., and J. N. Thompson. 1982. Phenology and ecology of color in bird-dispersed fruits or why some fruits are red when they are "green." *Can. J. Bot.* (in press).

Wilson, D. S. 1980. *The natural selection of populations and communities.* Benjamin/Cummings, Menlo Park, CA.

Wilson, E. O. 1971. *The insect societies.* Harvard University Press, Cambridge, MA.

Wilson, E. O. 1975. *Sociobiology: the new synthesis.* Harvard University Press, Cambridge, MA.

Wilson, E. O. 1976a. Behavioral discretization and the number of castes in an ant species. *Behav. Ecol. Sociobiol.* **1**:141–154.

Wilson, E. O. 1976b. A social ethogram of the neotropical arboreal ant *Zacryptocerus varians. Anim. Behav.* **24**:354–363.

Wilson, E. O., and R. M. Fagen. 1974. On the estimation of total behavioral repertories in ants. *J. NY Entomol. Soc.* **82**:106–112.

Wong, J. T. -F. 1975. A co-evolution theory of the genetic code. *Proc. Natl. Acad. Sci. USA* 72, pp. 1909–1912.

Yonge, C. M. 1968. Living corals. *Proc. R. Soc. Lond. B* **169**:329–344.

Zimmerman, J. G. 1932. Uber die extrafloren Nektarien der Angiospermen. *Beih. Bot. Zentralbl.* **49**:99–196.

INDEX